Dictio[illegible]
CHEM[illegible]

ANAND DHINGRA

SIT-Reference Series

STERLING PUBLISHERS PRIVATE LIMITED
A-59, Okhla Industrial Area, Phase-II, New Delhi-110020.
Tel: 26387070, 26386209; Fax: 91-11-26383788
e-mail: sterlingpublishers@airtelmail.in
www.sterlingpublishers.com

Dictionary of Chemistry

ISBN 978 81 7359 123 5
Reprint 2001, 2004, 2006, 2008

PRINTED IN INDIA

Printed and Published by Sterling Publishers Pvt. Ltd., New Delhi-110020.

PREFACE

There is an abundance of textbooks on chemistry, but dictionaries are hard to come by. This book is an attempt to fill that void.

It is primarily meant to be a source of reference for those who have very little exposure to chemistry. The entries are thus made with the beginner in mind. This does not, of course, mean that those who have already got a background in the subject will not benefit from it. The dictionary will refresh their memories, improve the clarity of their concepts, and shed light on aspects which the text books may not have highlighted.

Addressed to students and professionals, this dictionary has been updated to include the latest advances in the scientific field.

A

ABSOLUTE : Not dependent on any other quantity.

ABSOLUTE CONFIGURATION : Representation of the absolute structure of a compound with respect to the arrangement of atoms and groups in space, used for naming organic compounds. The R, S convention is usually used for representing absolute configuration. It is based on priority of groups attached to chiral carbon atoms. Priority is determined by atomic mass. The decreasing order of priority of some important groups is I, Br, Cl, $-SO_3H$, $-OCOCH_3$, $-OCH_3$ -OH, $-NO_2$, $-NH_2$, COO CH_3, $-CONH_2$, $-COCH_3$, -CHO, $-CH_2$, OH, $-C_6H_5$, $-C_2H_5$ $-CH_3$, -H. The molecule is viewed with the group of lowest priority behind the chiral atom. If the arrangement of other groups in the decreasing priority is clockwise, configuration is R otherwise it is S. R and S configurations are not related with the dextro-and leavo-rotatory character of the optically active molecules.

ABSOLUTE TEMPERATURE : Temperature on the absolute or Kelvin scale. It is denoted by T and is related to the temperature on Celsius scale as $T = (t + 273.15)$

ABSOLUTE ZERO : The temperature at which any ideal gas would have zero volume, if cooled without phase change. It corresponds to – 273.15°C. However, all gases liquify much before reaching this temperature.

ABSORBANCE : Capability of different organic compounds to trap a photon of energy and get excited.

ABSORPTION : It is the phenomenon of solution of gases in liquids or when a gas permeates into the whole body of a solid (having porous structure) rather than its surface.

ABSORPTION INDICATOR : An indicator used for determining end point in precipitation titrations. Due to the difference in the nature of ions being titrated (the indicator may absorb a particular type of ion, at the end point equivalence giving some colour).

ABSORPTION SPECTRUM : *See spectrum.*

ABUNDANCE : Relative amount of the given element with respect to others, usually expressed as a percentage; e.g., the abundance of oxygen in earth's crust is nearly 50%. It also refers to the relative amount of an isotope in the naturally occurring element, e.g., chlorine contains about 75% ^{35}Cl.

ACCELERATOR : A substance which may accelerate the rate of a reaction. It is also referred to a device used to accelerate particles like neutrons, alpha particles, in a nuclear reaction.

ACCEPTOR : An atom or group of atoms which may accept a one pair of electrons to form a coordinate (dative) bond. Such compounds are known as Lewis acids.

ACCUMULATOR : An electric cell or battery which may be recharged by passing an electric current in opposite direction from an external source.

ACETAL : An organic compound formed by the addition of alcohol to an aldehyde in the presence of dry HCl gas. The addition first gives hemiacetal (RCH (OH) OR') which changes into acetal $(RCH (OR')_2$. Acetals in the presence of excess alcohol can be decomposed by dilute acids to regenerate aldehydes.

ACETALDEHYDE : *See ethanal*

ACETAMIDE : *See ethanamide*

ACETIC ACID : *See ethanoic Acid*

ACETOACETIC ACID : CH_3COCH_2COOH. A colourless, syrupy keto acid which decomposes below 100° C to acetone and CO_2. Present in abundance in the urine of diabetic patients. Exhibits tautomerism and exists as a mixture of keto and enol forms.

ACETOACETIC ESTER : *See Ethylacetoacetate*, $CH_3COCH_2COOC_2H_5$. Belongs to the ester class. Colourless liquid with a pleasant smell. Prepared by Claisen's condensation i.e, the condensation of ethylacetate by sodium ethoxide. With dilute alkali it forms ketones used for the synthesis of large number of organic compounds. Shows tautomerism; keto form dominates.

ACETOL : CH_3COCH_2OH; Hydroxyacetone; Pyruvic alcohol. Colourless pleasant smelling liquid; reduces ammonical silver nitrate.

ACETOLYSIS : Reaction by which acetyl group of an organic compound is removed by heating with aqueous or alcoholic alkalies.

ACETONE : *See Propanone.*

ACETONE DICARBOXYLIC ACID : $CO(CH_2COOH)_2$; Keto glutaric acid. Colourless cystalline compound; used in organic synthesis.

ACETONYL : CH_3COCH_2 a radical.

ACETOPHENONE : $C_6H_5COCH_3$. Colourless plates with odour like almonds.

ACETYLATION : *See acylation.*

ACETYL CHLORIDE : *See ethanoyl chloride.*

ACETYLENE : *See ethyne.*

ACETYL GROUP : *See ethanoyl group.*

ACETYLIDE : *See carbide.*

ACHESON PROCESS : An industrial method developed by Edward Goodrich Acheson for manufacture of graphite by heating silicon carbide at a very high temperature of about 4420K.

ACHIRAL : A molecule which contains no chiral centre and hence optically inactive.

ACID : **Arrhenius theory** : An acid is a substance which increases hydronium ion concentration (H_3O^-) in aqueous solutions. H_3O^+ ion concentration in solutions should be more than 1×10^{-7} or pH should be below 7.9

Lowry-Bronsted theory : acid is a substance which may easily donate protons e.g, H_2O, HCO_3^-, HSO_4^-, H_2SO_4, CH_3COOH, etc. If a proton is lost, an acid forms its conjugate bases and vice versa.

Lewis theory : is based on the electronic theory which states that an acid is a substance which is electron deficient and has a tendency to gain a pair of electrons to form a dative bond, e.g. H_3O^+, BCl_3, $AlCl_3$.

Acid Anhydride : $(RCO)_2O$; Acid anhydride is prepared by the action of an acyl halide and sodium salt of a carboxylic acid. May also be prepared by heating with phosphorous pentoxide. They produce carboxylic acids on hydrolysis. They are also used for acylation purposes like acyl chloride.

Acid-Base Indicator : Aromatic compounds which exhibit change in colour within a narrow range of change of pH. Used in titrations to determine end point. Generally indicators act as weak acids or weak bases, e.g., phenolphthalein, methyl red, methyl orange, etc.

Acid Dyes : Dyes used in the dyeing of natural silk and wool from an acid dye bath containing dilute sulphuric acid or acetic acid. They are mostly salts of organic acids in which the chromophore is a negative ion, often an organic sulphonate. They may be compounds which produce acids when dissolved in water.

Acidic Hydrogen : Hydrogen atom which may be dissociated in solvent, e.g., acetylenic hydrogen, carboxylic hydrogen, phenolic hydrogen.

Acidimetry : Volumetric analysis in which a stanaard solution of an acid is titrated with a base of unknown strength using a suitable indicator. The strength of the base is thus determined.

Acidity Constant : *See dissociation constant.*

Acid Rain : *See pollution.*

Acid Salt : A salt in which all the replaceable hydrogen atoms of acid have not been replaced by metal cations or other cations. Usually they are formed by the partial neutralisation of polybasic acids, e.g., $NaHSO_4$, $NaHCO_3$, Na_2HPO_4.

Acid Value : It is the measure of the free acid present in oils, resins etc. It is defined as the number of milligrams of KOH required to neutralise free acid content of one gram of the substance.

Acrilan : A trade name of synthetic fibre polypropenonitrile.

Acrylic Acid : *See propenoic acid*

ACRYLIC RESINS : Resins made by the condensation polymerisation of esters of acrylic acid e.g, polypropenonitrile (acrilan), polymethylmethacrylate (PMMA).

ACTINIDES : Second row of f-block elements starting from Ac. Electrons successively fill in their 5f orbital. Elements $_{90}Th$, $_{91}Pa$ and $_{92}U$ occur in nature. Other higher elements are synthesised in the laboratory and are called transuranic elements. They are all radioactive and so the study of their properties is quite difficult.

ACTINIUM : A radioactive element. First member of the actinides. It has no electron present in 5f but due to the similarity in properties, it is studied along with actinides. It may be obtained by neutron bombardment of radium. It is used as a source of alpha particles. The most stable isotope, i.e., $_{89}Ac$ has mass 227 and half life of 21.6 years.

ACTINIUM SERIES : *See radioactive series.*

ACTINOMETER : An instrument to measure the intensity of electromagnetic radiation. It is based on the photoelectric effect.

ACTIVATED ALUMINA : *See aluminium hydroxide.*

ACTIVATED CHARCOAL : *See charcoal.*

ACTIVATED COMPLEX : The transition state in a chemical reaction at threshold energy barrier. In this state the atoms are partially bonded to one another.

ACTIVATION ENERGY : E_{act}. The minimum amount of energy reacting molecules must possess to react. It is the difference between the energy of the reactants and the threshold energy. It is represented in joules per mole of reactants.

ACTIVE MASS : *See (law of) mass action.*

ACTIVE SITE : A site on the surface of a catalyst which is more active due to the availability of free atoms. For enzyme molecules an active site is that which binds the substrate molecule. Also refers to that portion of a growing ion or radical through which polymerisation takes place.

ACTIVITY : **a**; Effective concentration. A thermodynamic property used in place of molar concentration in determining equilibrium constant for reactions involving non ideal gases and solution.

Activity coefficients for gases are given by a/p where p is pressure or by a / x where x is the mole fraction. Equilibrium constants K_c and K_p for reaction A = B+C may thus be shown as :

$$K_c = \frac{a_b a_c}{a_A}$$

ACTIVITY COEFFICIENT : *See activity and fugacity.*

ACYCLIC : A compound which is not cyclic which does not possess a closed ring.

ACYLATION : The reaction whereby an acyl group (R-C-group) is introduced in the benzene ring; or alcohols or amines; usually one H-atom is substituted by the group. Acylation is carried out by acyl halides (RCOX) or by acid anhydrides (RCOOCOR). In acetylation, the group introduced is CH_3CO-while for benzoylation the group introduced is C_6H_5CO-.

ACYL ANHYDRIDE : *See acid anhydride*

ACYL GROUP : RCO group.

ACYL HALIDE : RCOX. Can be prepared by the action of carboxylic acids with PCl_5, PCl_3 or $SOCl_2$. Acyl halides fume in moist air as they form HCl. They attack the hydrogen atom of alcoholic or amino groups. In IUPAC system, they are called alkanoyl halides.

ADDITION REACTION : A reaction in which two or more molecules combine. Found in the compounds having double or triple bonds, i.e., alkenes, alkynes, aldehydes, ketones, etc. They may be electrophilic (in alkenes, alkynes, arenes) or may be nucleophilic (in aldehydes, ketones alkynes).

ADDITION POLYMERISATION : *See polymerisation.*

ADDUCT : An addition product. The term is particularly used for compounds formed by the coordination between a Lewis acid and a Lewis base.

ADENINE : A purine derivative. It is one of the important component bases of nucleotides and the nucleic acids - DNA and RNA.

ADENOSINE : A nucleotide made by one adenine molecule joined to a ribose sugar molecule.

ADENOSINE DIPHOSPHATE : *See ATP*

ADENOSINE MONOPHOSPHATE : *See ATP*

ADENOSINE TRIPHOSPHATE : *See ATP.*

ADIABATIC CHANGE : A change taking place in an insulated vessel so that no heat enters or leaves the system. In an adiabatic process when work is done by the gas, volume increases and the temperature falls. [Under adiabatic conditions PV = const. (i.e, $\gamma = Cp/C_v$; Cp and C_v are heat capacities at constant pressure and at constant volume)]

ADIPIC ACID : *See hexandioic acid.*

ADRENALINE : Also called epinephrine. A hormone produced by the adrenal glands. It increases the heart activity, activity of muscles and the rate and depth of breathing. Simultaneously it checks digestion and excretion .

ADSORBATE : A substance which is adsorbed on the surface.

ADSORBENT : A substance on the surface of which molecules of adsorbate may be accommodated.

ADSORPTION : A process in which a layer of atoms or molecules of a substance (usually a gas – adsorbate) is formed on the surface of another substance (usually a solid – adsorbent). It may be of two types. Physical adsorption (physiosorption) is the adsorption where two types of molecules are held by weak Van der Waals forces of attraction. There may be several layers of adsorbate on the adsorbent. Chemical adsorption (chemisorption) is a process in which two types of molecules or atoms are held by strong covalent bonds. [In this case only one layer may be held by the surface.]

AEROSOL : A colloidal solution of solid or liquid in a gas (or air). Chlorofluorocarbons generally form aerosols but their use is discouraged on the ground that they damage the ozone layer.

AGAR : An extract of red seaweeds used as gelling agent in culture media, foodstuff, medicine, and cosmetic creams. Nutrient agar is made from beef extract or blood that is gelled with agar and used for the cultivation of bacteria, fungi, and algae.

AGATE : A variety of chalcedony which forms in rock cavities and has a pattern of concentric bonds or layers parallel to the cavity walls. Moss agate does not show the same pattern and is a milky chalcedony formed by inclusion of manganese and iron oxides. Agates are used in jewellery and for ornamental purposes.

AIR : Mixture of gases surrounding the earth. Composition of dry air by volume is : nitrogen 78.08%; oxygen 20.95%; argon 0.93%; carbon dioxide 0.03%; and neon, helium, krypton and xenon in still smaller amounts. It contains varying amounts of water vapour, dust, pollen, carbon particles, etc.

AIR GAS : *See producer gas.*

ALABASTER : A mineral form of gypsum.

ALANINE : *See amino acids.*

ALBUMIN : (Albumen) - Globular proteins soluble in water but which forms insoluble coagulates on heating. It occurs in blood, milk, plants and white of egg. Serum albumins, 55% of blood plasma proteins help regulate the osmotic pressure. Lactalbumin is a milk protein.

ALCOHOLS : Organic compounds containing –OH functional group. General formula : $(C_nH_{2n+1})OH$. In IUPAC nomenclature their names end in suffix - ol. When -OH group is attached to an aromatic ring the compounds are called phenols.

On the basis of number of -OH groups they may be monohydric alcohols (CH_3OH, CH_3CH_2OH etc); dihydric alcohols ($HO\ CH_2\ CH_2\ OH$, ethylene glycol); trihydric alcohols ($CH_2(OH)CH(OH)CH_2(OH)$, glycerol); and polyhydric alcohols CH2(OH)(CHOH)4CH_2OH, sorbitol). Based on the environment of C-OH grouping they may be classified as primary, secondary, and tertiary alcohols.

Primary alcohols are those where the -OH group is attached to a terminal carbon atom.

Secondary alcohols are those where the -OH group is attached to a carbon atom to two other carbon atoms.

Tertiary alcohols are those where the -OH group is attached to a carbon atom bonded to three other carbon atoms.

Alcohols in general are prepared by the hydrolysis of alkyl halides, controlled reduction of aldehydes, ketones and acids (by $LiAlH_4$ or by sodium amalgam in water), hydrolysis of esters, action of nitrous on acid, primary amines, etc. Among the main properties worth mentioning are - their oxidation to aldehydes and carbioxylic acids, dehydration to produce alkenes and aldalides, production of esters etc.

ALDEHYDES : Group of organic compounds having general formula $C_n H_{2n+1}CHO$. Structurally aldehyde group is $-CH=O$. In IUPAC systems the names of aldehydes end in - al. Examples may be formaldehyde (HCHO, methanal); acetaldehyde (CH_3CHO, ethanal).

Aldehydes are formed by oxidising the primary alcohols by potassium dichromate or potassium permanganate and sulphuric acid at the boiling temperature of aldehyde. They may be obtained by the dehydrogenation of primary alcohol by heated copper. Aromatic aldehydes are made by the action of HCl and CO in the presence of anhydrous $AlCl_3$ (Gattermann -Koch synthesis); by the action of chromyl chloride (Etard's reaction) or by the action of chloroform on alkaline phenol solution (Reimer - Tiemann reaction).

Aldehydes undergo nucleophilic addition reactions and form several important compounds - cyanohydrins with hydrogen cyanide; aldoximes with hydroxylamine; acetals with alcohols; etc. They undergo aldol condensation (e.g, CH_3CHO); Perkin's and benzoin condensation (C_6H_5 CHO); Cannizzaro's reaction (HCHO and C_6H_5CHO) etc.

Aldehydes are very easily oxidised to carboxylic acids and so act as reducing agents. They reduce Tollen's reagent, Fehling solution, etc. They are distinguished from other compounds by silver mirror test, Fehling solution test and Schiff's test.

ALDOHEXOSE : A sugar having six carbon atoms including one aldehyde group. *See sugar.*

ALDOL : Aldhydes having at least one hydrogen atom, condense in presence of base to make hydroxyaldehyde called aldol. Presence of α-H atom helps in making the nucleophile with a base carbanion. The nucleophile may then attack another molecule of same or different aldehyde.

Aldol thus formed when heated with an acid, loses water and produces unsaturated aldehyde.

ALDOPENTOSE : An aldose having five carbon atoms. *See sugar.*

ALDOSE : A sugar which has an aldehyde group in free state or in the ring form. *See sugar.*

ALDOSTERONE : Hormone produced by the adrenal glands which regulates the excretion of sodium by the kidneys and hence maintains the balance of salt and water in the body.

ALDOXIMES : Compounds with general formula RCH=N-OH, obtained by the reaction between an aldehyde and hydroxyl amine. Aromatic aldoximes are crystalline solids. On heating with dilute mineral acids, they regenerate aldehyde. All aldoximes exist in two forms called syn and anti, depending on the different orientation of groups.

ALDRIN : An insecticide. It is toxic to human and animals.

ALICYCLIC : Cyclic aliphatic compunds such as cyclohexane, cyclopentane, etc.

ALIPHATIC COMPOUNDS : Organic compounds having carbon atoms in open chains. They are alkanes, alkenes, and their derivatives. Some of these compounds have closed chains also. In these cases the double bonds are not conjugated unlike aromatic compounds where electrons are delocalised about the ring.

ALIZARIN : It is an orange red dye soluble in organic solvents. Its structure has three fused benzene rings.

It is prepared by fusing anthraquinone - 2- sulphonic acid with NaOH and little $KClO_3$. It forms a red lake with $Cr(OH)_3$ and a violet lake with $Fe(OH)_3$. It is dissolved in boiling water for dyeing.

ALKALI : Water soluble strong base. Generally refers to hydroxides of alkali metals (like Na, K, etc). Hydroxides of other metals also give alkaline solutions.

ALKALINE : An aqueous solution whose pH is more than 7.

ALKALI METALS : Elements of first group of periodic table. These are Li,Na, K, Rb, Cs and Fr; last of these is radioactive metal. These elements are highly electropositive. They form ionic compounds. All have the same valence shell configuration ns^1 where n is the outermost orbit. All have lowest ionisation energies and largest sizes in their respective periods. Within the groups, the elements exhibit an increase in size and decrease in ionisation energies.

All these elements react with water, air, halogens, sulphur and hydrogen. Carbonates and hydroxides of alkali metals are thermally stable and do not decompose on heating. Nitrates of alkali metals decompose to give the corresponding nitrate and oxygen.

Lithium differs from the rest of the alkali metals due to its smallest size and maximum charge to size ratio. It is diagonally related to magnesium of the second group. Lithium hydroxide carbonate and nitrate decompose on heating to form oxides. Alkali metals also form organometallic compounds. These compounds, particularly the lithium compounds, are widely used to synthesise a large number of organic compounds.

ALKALIMETRY : Volumetric analysis to determine the strength of acids.

ALKALINE EARTH METALS : Elements of second group of periodic table. These are Be, Mg, Ca, Sr, Ba and Ra. These elements are moderately reactive metals, harder than the alkali metals. The term **alkaline earth** refers to the oxides but is often used for the elements themselves. Valence shell configuration of these elements is ns^2. The group shows increasing metallic trend from Be downwards. Be, due to its smallest size and maximum charge to size ratio, exhibits high polarising power and forms covalent bonds. Berryllium hydroxide is amphoteric in nature while hydroxides of other elements are basic in nature with increasing base strength and solubility. Solubility of sulphates on the other hand shows a decreasing trend.

All alkaline earth elements burn in air to give oxides except barium which forms peroxide.

These oxides are soluble in water except that of Be. Metals except Be directly reacts with water to liberate H_2. A similar trend is also observed in the reaction of hydrogen. Be does not form any hydride directly.

Magnesium forms organometallic compounds called Grignard's reagents with general formula RMgX. These are widely used in organic sythesis.

Berryllium shows diagonal relationship with aluminium of third group.

All the isotopes of radium are radioactive.

ALKALOID : A group of compounds found in plants. Examples are morphine, codeine, caffeine and nicotine. They contain oxygen and nitrogen besides carbon and hydrogen.

ALKANALS : Organic aldehydes.

ALKANES : Saturated hydrocarbons called paraffins with general formula C_nH_{2n+2}. Alkanes are quite unreactive by nature. They undergo substitution reactions like halogenation, nitration and sulphonation. Most important sources of alkanes are natural gas (mainly methane) and crude oil (mixture of hydrocarbons). There are several methods for preparing alkanes in the laboratory : decarboxylation of carboxylic acids by soda lime or by electrolysis; reduction of halides by HI or by zinc copper couple and alcohol; action of water on Grignard's reagent; action of sodium metal with alkyl halide (called Wurtz reaction).

Alkanes burn in air to produce large amounts of energy and hence used as fuels.

ALKENES : Unsaturated hydrocarbons with the formula C_nH_{2n} primarily called alkenes have a carbon-carbon double bond. Examples are ethene (C_2H_4), propene (C_3H_6), etc. They are prepared by the dehydration of alcohols by conc. sulphuric acid, phosphoric acid or alumina; by the dehydrohalogenation (elimination) of alkyl halides by alcoholic caustic potash; by the dehalogenation of vicinal halidas using zinc as a catalyst. Alkenes undergo addition reactions when they take up a molecule of halogen to form vicinal dihalide, a molecule of halogen acid to form alkyl halide and a molecule of water to form alcohol.

Alkenes are distinguished from alkanes by Baeyer's test and bromine water test. When treated with alkenes, alkaline $KMnO_4$ and bromine water are decolourised.

Alkenes undergo chain reactions to form polymers. For this high temperature and high pressure is needed. Some catalysts may also be used. Examples of such polymers are polythene, polypropylene, etc.

Alkenes undergo ozonolysis, through which aldehydes and ketones can be prepared. This reaction is used to locate the position of double bond in alkenes.

ALKOXIDE : Compounds having RO group where R is an alkyl group. These compounds are prepared by the action of metals like Na and K on alcohols such as sodium methoxide CH_3ONa.

ALKYLBENZENE : An aromatic hydrocarbon containing one of more alkyl groups substituted in place of H atoms in the benzene ring. Examples are toluene or methylbenzene ($C_6H_5CH_3$) and xylenes. These compounds are obtained by Friedel Crafts reaction or by Wurtz-Fitting's reaction.

Alkyl group on benzene ring is an activating and directing group towards electrophilic substitutions.

ALKYL GROUP : A group obtained by the removal of a H-atom from an alkane. Examples are methyl ($-CH_3$); ethyl ($-CH_2CH_3$) groups.

ALKYL HALIDE : *See haloalkanes*.

ALKYNE : Unsaturated hydrocarbons containing carbon - carbon triple bond. Their general formula is $C_n H_{2n-2}$. In IUPAC system, they are represented by suffix-*yne*. Examples are ethyne (HC-CH); propyne (CH_3-C=CH). The simplest and most important member **ethyne** (acetylene) is prepared by the action of water on calcium carbide. In general, alkynes are prepared by the action of hot alcoholic potash solution on dihaloalkanes or by the action of zinc dust on tetrahaloalkanes. Like alkenes, the alkynes also undergo addition reaction with H_2, halogen, and halogen acids. It reacts with mercury (II) ions form ketones (ethyne gives ethanal).

The H-atom directly attached to the triple bonded carbon atom can be replaced and so terminal alkynes are acidic in nature. They form precipitates with ammonical $AgNO_3$ and ammonical Cu_2Cl_2.

ALLOTROPY : The capability of elements to exist in two or more physically different forms. Such forms, called allotropes, show similar chemical properties. Most common examples are the elements of groups IVA, VA and VIA of the periodic table. Important allotropes of carbon are diamond and graphite. Sulphur exists in rhombic and monoclinic forms while phosphorous shows three forms—white, red and black.

ALLOY : A homogeneous mixture of two or more metals or a metal with non metals. Examples are brass (alloy of Cu and Zn), steel (alloy of Fe, C and some other metals like Cr, Ni, Mn). Sometimes they may contain small particles of one phase in another phase. Alloys are made to enhance existing properties in a particular metal, e.g steel is made to increase hardness, strength and resistance to corrosion.

ALLYL GROUP : A group with formula $CH_2=CH\text{-}CH_2$. Also called propylene group.

ALLYL POLYMERS : Polymers derived from monomers containing an allyl group. Allyl phosphates make flame-resistant polymers. Allyl ethers of carbohydrates mixed with styrene and other vinyl compounds form polymers used as adhesives.

ALNICO : A group of alloys of iron containing aluminium, nickel, cobalt and copper. They are used to make powerful permanent magnets.

ALPHA PARTICLE- : A He^2+ ion emitted by the radioactive decay of large unstable nuclei. Alpha decay causes the formation of new nuclei having mass number less by four units and atomic less by two units. A stream of alpha particles is called alpha rays or alpha radiation. Alpha particles are used in artificial radioactivity. Alpha particles are deflected towards a negative field. They can ionise the gas through which they move and can influence photographic films.

ALUM : A type of double sulphates obtained by crystallising mixtures from the aqueous solutions of sulphates in proper proportions. Their general formula is M_2SO_4. $M_2(SO_4)_3.24H_2O$ where M is a monovalent metal ion or a group while M is a trivalent metal ion. Aluminium potassium sulphate (called potash alum), i.e. K_2SO_4, $Al_2(SO_4)_3 24\ H_2O$ is a common example. Some other alums are ammonium alum-$(NH_4)_2\ SO_4$. $Al_2\ (SO_4)_3$. $24H_2O$, chrome alum -($K_2\ SO_4$, $Cr_2\ (SO_4)_3$. 24 H_2 O - Ferric alum) $(NH_4)_2\ SO_4$. $Fe_2\ (SO_2)_3$. 24 H_2O. Common potash alum is used for purifying water, as a mordant for dyes and as a water purifying agent.

ALUMINA Aluminium oxide, Al_2O_3. It is a white powdery substance, insoluble in water. It is used to make aluminium metal by electrolysis. Alumina occurs naturally as bauxite, corundum and diaspose. It is used in the furnace linings due to its refractory nature. Alumina acts as a dehydrant and a catalyst in several organic reactions.

ALUMINATES : *See aluminium hydroxide.*

ALUMINIUM : Second member of 13th group of the periodic table. Its atomic number is 13 and r.a.m. is 26.97. It is a soft and moderately reactive metal. It is the most common metal in the earth's crust (81% by weight). Commercially important ores are bauxite (Al_2O_3), cryolite (Na_3AlF_6) and mica (Aluminosilicate). It is produced commercially by Hall-Heroult process in which concentrated alumina, mixed with cryolite, is electrolysed in molten state.

Aluminium atom is much bigger in size than that of boron. It may form Al^{3+} ions, particularly in aqueous solutions. It shows some non-metallic properties also and forms a number of covalent compounds.

It forms only a few hydrides-AlH_3 and Al_2H_6. It forms a mixed hydride - $LiAlH_4$ which is used as an important reducing agent in organic reactions.

Aluminium reacts with oxygen, liberating heat but at ordinary temperatures, a protective layer of oxide is formed. Aluminium reacts with acids forming ions while with alkalies it forms aluminates $[Al(OH)_4]$.

Aluminium reacts readily with halogen at suitable temperatures. AlF_3 is ionic while others are dimeric covalent compounds in vapour phase. Aluminium forms nitride and carbide (Al_4C_3) at very high temperatures.

It may form various alloys with Cu, Mn, Si, Zn and Mg. Due to its low density, strength of alloys, corrosion resistance and good electrical conductance, it is used for vehicle parts, aircraft components, window and door frames, and power cables.

ALUMINIUM ACETATE : $Al(COOCH_3)_3$, aluminium ethanoate is a white solid, slightly soluble in water. It can be prepared by reacting aluminium chloride and acetic anhydride at 180°C. It is usually obtained as the dibasic salt $Al(OH)(COOCH_3)_2$ (called basic aluminium acetate) by treating aluminium hydroxide with acetic acid. It is used in dyeing as a mordant, in paper industry for sizing, hardening, and in tanning.

ALUMINIUM BROMIDE : $AlBr_3$; A white solid soluble in water and several organic solvents.

ALUMINIUM CHLORIDE : $AlCl_3$; A solid which fumes in moist air due to formation of hydrogen chloride; reacts violently with water. It is prepared by passing chlorine or hydrogen chloride gas over hot aluminium or by passing chlorine over hot aluminium oxide and carbon. In liquid and vapour phases, it exists in its dimeric state, Al_2 Cl_6, in which two Al atoms are bridged together through two chlorine atoms forming dative bonds.

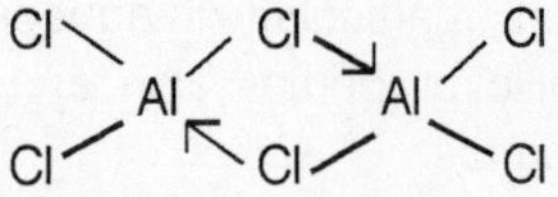

```
Cl\     /Cl\     /Cl
    Al       Al
Cl/    \Cl/     \Cl
```

It exists as hexahydrate which can not be converted into anhydrous salt by heating. On heating it changes into aluminium oxide.

Aluminium chloride is used commonly as catalyst for cracking of oils. It is also used as a catalyst in many organic reactions, e.g., Friedal Crafts reaction.

ALUMINIUM HYDROXIDE : $Al(OH)_3$; A white crystalline solid. The compound occurs in nature as a mineral gibbsite. It is obtained by bubbling hydrogen sulphide or carbon dioxide gas through the aqueous solution of some aluminium salt.

It has coordinated water molecules and therefore also known as hydrated aluminium hydroxide. It is amphoteric in nature. In strongly basic medium it forms aluminate while in the presence of mineral acids, it forms corresponding salts.

On heating, the hydroxide transforms to a mixed oxide hydroxide. On strong heating it changes to various activated forms of alumina. These forms differ in porosity; these are used as catalysts and as an adsorbent in chromatography. Gelatinous freshly precipitated aluminium hydroxide is used as mordant for dyeing because of its tendency to make coloured lakes with organic dyes.

ALUMINIUM NITRATE : $Al(NO_3)_3.9H_20$; A hydrated white crystalline solid. It is made by dissolving aluminium hydroxide in nitric acid.

ALUMINIUM OXIDE : Al_2O_3. *See alumina*.

ALUMINIUM POTASSIUM SULPHATE : Potash alum; K_2SO_4 $Al(SO_4)_{324}$. H_2O. It is a double salt and can be prepared by crystallisation from a solution containing equimolar amounts of two components.

Potash alum loses eighteen molecules of water at about 366 K and becomes anhydrous at 473 K. *See alums also*.

ALUMINIUM SULPHATE : Al_2 $(SO_4)_3.18H_20$. It occurs in the mineral alunogerite. It loses water at 359.5K. It may be prepared by dissolving aluminium hydroxide or china clay in sulphuric acid.

The anhydrous form decomposes on strong heating to form alumina, sulphur dioxide and sulphur trioxide.

Its aqueous solution is acidic due to cationic hydrolysis. It is used in sewage treatment as a coagulating agent and in the purification of drinking water.

ALUMINOSILICATES : Silicates containing Al. Examples are feld-spar, mica, zeolitic, clay, etc.

ALUNDUM : An artificial variety of corrundum (Al_2O_3). It is formed by fusing bauxite in an electric furnace and allowing the molten product to cool rapidly. It is used to make refractory bricks, crucibles and muffle furnaces.

ALUNITE : Also called alumstone. It is used as a commercial source of potash alum and aluminium sulphate.

ALUNOGERITE : A mineral form of aluminium sulphate.

AMALGAM : An alloy of mercury with one or more metals. It may be liquid or solid. Iron and platinum cannot form amalgams. An amalgam of sodium with water is used as a source of nascent hydrogen while zinc amalgam with concentrated hydrochloric acid is used as a reducing agent in Clemmenson's reduction.

AMARANTH : A red dye used in foodstuff.

AMATOL : A mixture of ammonium nitrate and trinitrotoluene used as explosive.

AMBIDENT : Ligands which can use two coordination sites.

AMBLYGORITE : $LiAlF(OH)PO_4$. An ore of lithium.

AMERICIUM : Atomic number 95. Mass number of stable isotope is 243. It is a member of actinide series (5f series). It is a transuranic element, synthesised from plutonium. G.T. Seaborg synthesised it for the first time by bombarding U with alpha particles.

AMETHYST : A purple variety of the mineral quartz. The colour is due to impurities like iron oxide. It is used as a gemstone.

AMIDE : A type of organic compound of general formula $RCONH_2$. Amides are white crystalline solids, basic in nature; some of these are water soluble. They are made by the ammonolysis of acid derivatives. They are also obtained by heating ammonium salt of a carboxylic acid. Amides, on heating with phosphorous pentoxide, form nitriles. On hydrolysis, amides form carboxylic acids.

Amides on heating with bromide and alcoholic potash solution change into amines having one carbon atom less. (*See Hofmann degradation*)

Inorganic salts containing - NH_2 group such as $NaNH_2$ and KNH_2. They are formed by the action of ammonia with certain metals like alkali and alkaline earth metals (expects Be and Mg).

AMINATION : A reaction in which an amino group is introduced into a molecule. Example maybe the formation of amine by the reaction of alkyl halide with ammonia under suitable conditions of pressure and temperature.

AMINES : A group of compounds obtained by replacing one or more hydrogen atoms of ammonia by alkyl groups.

Amines are classified according to the number of alkyl groups attached to the nitrogen atom, i.e., RNH_2 - primary, R_2NH secondary and R_3N tertiary amines. They are all basic in nature. All of them along with the quaternary ammonium salt $[R_4N]^+x$ can be prepared by Hofmann's ammonolysis of alkyl halides.

They react with acids to form salts. With acid chlorides they form substituted amides (RCONHR and $RCONR_2$).

AMINE SALTS : Ammonium salts in which one or more H – atoms are replaced by alkyl groups. Amines form such salts by reacting with acids, common examples being ethyl ammonium chloride $C_2H_5NH_3{}^+Cl$, dimethyl ammonium chloride $C_6H_5NH_3{}^+Cl$. These salts are crystalline substances. These are dissolved in water readily. On addition of caustic alkali, free amine is liberated.

When all the four H-atoms of an ammonium salt are substituted, quaternary salts are produced eg. $(CH_3)N + Cl$. Such salts do not liberate amine on reacting with caustic alkali but form quaternary ammonium hydroxides which are strong bases.

AMINO ACIDS : Amino derivatives of aliphatic carboxylic acids eg. glycine NH_2-CH_2-CO-OH. They are white crystalline water soluble compounds. Except glycine, all alpha-amino acids are optically active. Most important amino acids are synthesised by the body, these are called non essential amino acids. Remaining are essential and have to be supplied in the diet.

Alpha amino acids have complex formulae and are usually represented by common names eg. alamine $CH_3CH(NH_2)COOH$; cystine $[SCH_2\text{-}CH(NH_2)COOH$; glutamic acid $COOH\ (CH_2)_2\ CH\ (NH_2)COOH$; histidine $C_3H_3N_2CH_2\ (NH_2)\ COOH$; lysine $NH_2\ (CH_2)Cl(NH_2)\ COOH$.

Through the formation of peptide linkages, amino acids join one another to form peptides (short chains). or polypeptides (much longer chains). Polypeptides with very high molecular masses (more than 10,00) are called proteins. The sequence of amino acids in the proteins determines their shape, properties and biological role. Plants and some microorganisms may synthesise amino acids from simple inorganic substances, but animals depend on their diet for these.

AMINO AZOBENZENE : $O\text{-}N=N\text{-}O\text{-}NH_2$ prepared by the action of aniline and benzene diazonium chloride in ice cold, slightly acidic medium. Used as a first component in the preparation of azo dyes.

AMINOBENZENE : *See aniline.*

AMINOETHANE : *See ethylamine.*

AMINO GROUP : NH_2 group. *See amines also.*

AMINE : NH_3 molecules in the inorganic complex are called amine molecules; the coordination complexes in which the ligands are ammonia, e.g., $[CN(NH_3)_4]SO_4$.

AMMONIA : NH_3; A colourless, pungent smelling gas. Condenses to colourless liquid and ultimately to a white solid. It is lightly soluble in water. It is present in the atmosphere in very small amounts. It may be prepared in the laboratory by heating an ammonium salt with a strong base or by the hydrolysis of metal nitrides. Industrially it is manufactured from nitrogen and hydrogen gases by Haber's process. Normally it does not burn in air but may react with atmospheric oxygen at high temperature in the presence of a heavy metal catalyst. Product of this reaction is nitric oxide which is further used for manufacturing nitric acid.

Ammonia reacts with acids to form salts and with active metals like Na, K, it forms amides. Ammonia is also a reducing agent (reduces halogens). It is also an important laboratory reagent.

Liquid Ammonia is an important non-aqueous solvent which has some similarity with water. Like water, its molecules remain in the associated state due to intermolecular hydrogen bonding. It has moderate dielectric constant ions.

Alkali metals and alkaline earth metals (except Ba and Mg) dissolve in liquid ammonia to give blue coloured, conducting solution which acts as strong reducing agents. It is available in the market as liquid, compressed in the cylinders or as aqueous solutions of various strengths.

AMMONICAL : Describing a solution which has aqueous ammonia as a solvent.

AMMONIUM ALUM : Alum containing ammonium sulphate as a component.

AMMONIUM BICARBONATE : NH_4HCO_3; A white crystalline compound prepared by passing excess amount of carbon dioxide in aqueous ammonia solution. It is used in medicines.

AMMONIA BROMIDE : NH_4Br; A coloured crystalline substance soluble in water. Readily sublimes on heating.

AMMONIUM CARBONATE : $(NH_4)\ CO_3$; A white or colourless substance which crystallises as plates, usually exists as monohydrate. It is readily soluble in water and decomposes on heating to ammonia, carbon dioxide and water. The commercial aminomethanoate (NH_2CONH_4) is obtained by the action of ammonium chloride on calcium carbonate followed by sublimation. It decomposes to ammonium bicarbonate and ammonia. Commercial ammonium carbonate is used in baking process, smelling salts, in dyeing, and in wool industry.

AMMONIUM CHLORIDE : NH_4Cl; Salt of ammonia, a white crystalline solid, highly soluble in water, only slightly soluble in ethanol but insoluble in ether. It may be obtained by the reaction between gaseous ammonia and hydrogen chloride gas. It may also be prepared by fractional crystallisation from a solution of ammonium sulphate and sodium chloride or ammonium carbonate and calcium chloride.

It is used in dry cells, as a flux for soldering, in dyeing and in calicoprinting.

AMMONIUM CHROMATE : $(NH_4)_2CrO_4$; A golden yellow solid, soluble in water. Decomposes on heating to dichromate ammonia and water.

AMMONIUM DICHROMATE : $(NH_4)_2 Cr_2O_7$; A red crystalline solid. Soluble in water, decomposes on heating to nitrogen, water and chromic oxide.

AMMONIUM HYDROXIDE : NH_4OH; An aqueous solution of ammonia. It is a weak base and forms salts with acids.

AMMONIUM ION : NH_4^+; A monovalent cation formed by the linking of a proton with ammonia. It may be considered as a Bronsted acid of ammonia. It has a tetrahedral shape. Chemical properties are similar to those of alkali metal salts.

AMMONIUM MOLYBADE : $(NH_4)_6MO_7O_{24}$. $4H_2O$. Salt is used to test phosphatel anion as well arsenic cation in the laboratory.

AMMONIUM NITRATE : NH_4NO_3; A colourless crystalline solid, soluble in water and ethanol. Prepared by the action of nitric acid or ammonia. Because of high nitrogen content, it is widely used as a fertilizer. It is also used in the manufacture of explosives.

AMMONIUM SULPHATE : $(NH_4)_2SO_4$; A white crystalline solid, soluble in water. It occurs in nature as the mineral masagnite. It is prepared by the direct reaction between ammonia gas and sulphuric acid. On heating; it decomposes to ammonium hydrogen sulphate which on stronger heating liberates nitrogen, ammonia, sulphur dioxide and water. It is widely used as a fertilizer but its use tends to make the soil acidic.

AMMONIUM SULPHIDES : $(NH_4)_2S$ and NH_4HS are comparatively unstable. A solution of ammonium hydrosulphide is prepared by passing hydrogen sulphide gas through strong aqueous ammonia. It is called yellow ammonium sulphide and is used in qualitative analysis.

AMORPHOUS : A solid substance which is not crystalline. The solid which has no regular order of atoms or molecules in its lattice. Glasses are examples of this type. Recently, X-Ray analysis has shown that the substances which were previously described as amorphous are actually composed of very small crystals, e.g., charcoal, coke etc., are made up of small graphite-like crystals.

AMPERE : Symbol A; The SI unit of electric current, defined as the constant current that is maintained in two straight parallel infinite conductors of negligible cross section placed one metre apart in a vacuum producing a force of 2×10^{-7} newton per metre between the conductors.

Earlier it was defined as the current required to deposit 0.0011180 g of silver from a solution of silver nitrate in one second. It named after the name of scientist A.M. Ampere.

AMPERE HOUR : A practical unit of electric change equal to the charge flowing in one hour through a conductor. It equals 3600 coulombs.

AMPHIBOLES : A group of metasilicate minerals which make rocks. They are of double chain silicate types present in many igneous and metamorphic rocks. Examples include tremolite, $Ca_2\ Mg_5\ (SigO_{22})(OH,F)_2$; $Ca_2(MgFe)_5\ Si_8O_{22}\ (OH,F)_2$; edenite $Na\ Ca_2\ (Mg\ Fe)_5\ (S_{i7}\ Al\ O_{22})(OH,F)_2$.

AMPHOLYTE : A substance which can act as an acid in the presence of a strong acid, e.g., aminoacetic acid; sulphanilic acid.

AMOLYTE ION : *See zwitterion.*

AMPHOTERIC : A substance which shows both acidic as well as basic properties. It is generally used for oxides and hydroxides e.g. Al_2O_3; $Al\ (OH)_3$; ZnO; PbO; Pb_2; PbO_2 etc. Sometimes the existence of these oxides indicate the metalloidal nature of the element.

AMU : *See Atomic mass unit.*

AMYL ALCOHOL : $C_5 H_{11}OH$; An aliphatic alcohol.

AMYLASE (DISTASTE) : Enzyme which decomposes starch, glycogen and other polysaccharides. Plants contain both α and β amylases. Animals possess only α amylases secreted by pancreas and also present in human saliva. In general, they decompose polysaccharides to glucose and maltose.

AMYL GROUP : Any of the several isomeric C_5H_{11} groups.

AMYLOPECTIN : A polysaccharide consisting of lightly branched chains of glucose molecules. It is the water insoluble component of starch. *See starch also*.

AMYLOSE : A polysaccharide of linear chains of glucose molecules. It is the water soluble component of starch. It gives a characteristic blue colour with iodine.

ANABOLISM : All the metabolic reactions which synthesise complex molecules of proteins, fats and other vital constituents from simple molecules. This process needs energy in the form of ATP.

ANALYSIS : The method of determining the components or constituents in a given sample. Analyses is of two types-qualitative and quantitative (to measure the actual amounts of different components).

Quantitative analysis include volumetric, gravimetric, chromatographic, spectroscopic techniques, etc.

ANAESTHETIC : The substances responsible for the loss of bodily sensation. Local anesthetics may act at only the site of application by freezing (ethyl chloride) or by affecting the nerves (cocaine). General anaesthetics like chloroform produce total anaesthesia.

ANDREW'S EXPERIMENT : An experiment to establish a relation between the pressure and volume for a given mass of carbon dioxide at constant temperature. The isotherms plotted show clearly the existence of a critical point and leads to the understanding of the liquefaction of gases.

ANGLESITE : A mineral form of lead (II) sulphate $PbSO_4$.

ANGSTROM : A° . A unit of length equal to 10^{-10} metre. It is used for expressing wavelengths, intermolecular distances, sizes of molecules and bond distances. Now this unit is replaced by nanometre 1A°=0.10nm. This unit is named after the Swedish scientist A.J. Angstrom.

Anhydride : A compound formed by removing water from an acid. Non-metal oxides are anhydrides of acids e.g. SO_2 is the anhydride of sulphurous acid. *See acid anhydrides also.*

Anhydrite : An important rock containing anhydrous mineral form of calcium sulphate under natural conditions. It slowly changes to form gypsum. It is used as a raw material in the manufacture of cement and fertilizers.

Anhydrons : Salts when separated from their water of crystailisation change into anhydrous state e.g. $CuSO_4.5H_2O$, a blue crystalline compound changes into $CuSO_4$ anhydrous white mass.

Aniline : $C_6H_5NH_2$; aminobenzene; phenylamine. A colourless aromatic substance obtained by the reduction of nitrobenzene by tin and hydrochloric acid in laboratory. Develops red brown colour by aerial oxidation. It is used for making dyes, medicines and important organic compounds.

Animal Charcoal : *See charcoal.*

Animal Starch : *See glycogen.*

Anion : A negatively charged atom or group of atoms. It is generally formed by the gain of one or more electrons by atoms. During electrolysis, they get attracted to the anode. Examples are Cl^-, NO_3^-, OH^-, etc.

Anionic Detergents : *See detergents.*

Anionic Resins : *See ion exchange.*

Anisotropy : A phenomenon shown by certain substances whereby one or more physical properties differ in different directions. Crystals are generally anisotropic with respect to some physical properties like transmission of electromagnetic radiations.

Annealing : A type of heat treatment to alter the properties of metals, particularly the hardness, e.g., malleability and brittleness. In this process the metal is heated to a particular temperature for some specified time and then cooked slowly.

The temperature to which a metal is heated depends on the metal concerned and the properties to be introduced. It is applicable to metals as well as other materials like glass.

Anode : The electrode at which oxidation takes place. In electrolysis, anode is the positive electrode and attracts anions. In an electronic vacuum tube, it attracts electrons from cathode. In a galvanic cell, the anode is negative and emits electrons spontaneously.

Anode Sludge : *See electrolytic refining.*

Anodizing : A method of protecting aluminium objects with protective oxide layer by making them the anode in an electrolytic bath containing sulphuric acid or some other oxidising solution. It can also be used to produce a decorative film of anoxide which can absorb a coloured dye.

Anomes : A term used to express the stereoisomers of carbohydrates. In case of glucose, αD(+) glucose and β D(+) glucose are diastereomers which differ in configuration only around C_1. These forms have different physical properties like melting point and specific rotation.

Antacids : Substances used to reduce the amount of acid in the stomach. Besides milk of magnesia and sodium bicarbonate, some vegetables also serve as antacids.

Anthracene : A white crystalline compound with molecular formula $C_{14}H_{10}$. Structure has these benzene rings fused into one another. It is obtained by the fractional crystallisation of green oil component of crude oil. It is an aromatic compound having an electron cloud of fourteen delocalised π-electrons. It is extensively used in the manufacture of dyes.

Anthracite : A form of coal which contains nearly 93 percent carbon.

Antibiotic : An organic compound generated by micro-organisms. It is capable of checking the growth as well as destroying other harmful micro-organisms. Examples are penicillin, streptomycin, and tetracycline. Antibiotics are used to treat various infections but they tend to weaken the body's natural defences, thus causing allergies.

ANTIBODIES : Protein molecules present in the serum which are produced in the body in response to the foreign bodies causing diseases. The antibody-antigen mechanism is the basis of immunology

ANTIBONDING ORBITAL : *See orbital.*

ANTIFOAMING AGENT : Compounds which discourage the formation of foams, e.g., octanal.

ANTIFREEZING AGENT : A substance used to prevent the freezing of water (as coolent) in the motor engines e.g. ethylene glycol. It is usually mixed with water in European countries when the temperature falls much below 273 K.

ANTIGENS : Some micromolecular proteins, which on introduction into the blood of animals, stimulate the production of antibodies.

ANTIKNOCK AGENT : A compound added to petrol to check or reduce the extent of knocking in the engine e.g. tetraethyl lead.

ANTIMONIC : An antimony (IV) compound.

ANTIMONOUS : An antimony (III) compound.

ANTIMONY : Sb; A toxic metalloid belonging to group VA of the periodic table with atomic number 51 and relative atomic mass 121.75. It exists in several allotropic forms like yellow antimony, black antimony (both unstable), and bluish white antimony (the stable form). Important mineral of antimony is stibnite Sb_2S_3. The metal finds use as an alloying agent in lead accumulator plates, type metal, solder, Britannia metal, etc. It is used to make semiconductors with germanium of group IVA. Its compounds are used in paints, ceramics, glass dye stuff and rubber technology.

The element remains unaffected by air, water and dilute acids but may burn in air. It reacts with oxidising agents and halogens.

ANTIMONY (III) CHLORIDE : $SbCl_3$; A white deliquescent solid called butter of antimony. It is readily hydrolysed by water to make antimonyl chloride. SbOCl (also called antimony (III) chloride oxide).

ANTIMONY (III) OXIDE : Sb_2O_3; A white amphoteric solid substance. Insoluble in water. It has strong tendency to act as a base. It is made by the action of air, oxygen or steam on antimony. It is also formed by boiling antimony (III) chloride solution in water.

ANTIMONY OXIDE : Sb_2O_5; A yellow solid formed by the action of nitric acid on antimony or by the hydrolysis of antimony (V) chloride. It is acidic in nature but dissolves in water only slightly.

ANTIMONY POTASSIUM TARTRATE : A white crystalline solid used to induce vomitting and so it is called tartar emetic.

ANTIOXIDANTS : Compounds which decrease the oxidation rate in certain substances, e.g., aromatic amines and substituted phenols. Used to protect fats and preservatives.

ANTIPYRETICS : Medicines used to lower the body temperature in case of fever ego paracetamol.

ANTISEPTICS : Compounds which may check the growth of unwanted bacteria and micro-organisms if applied to living tissues, e.g., iodine.

ANTITOXINS : Antibodies formed in blood to which bacterial toxins are introduced. An antitoxin may neutralise its own toxin with which it apparently combines quantitively.

APATITE : $CaF_2(PO_4)_3$; A naturally occurring phosphate of calcium, the most common of the phosphate minerals. Occurs mostly with metamorphic rocks especially limestone. It is used in the manufacture of fertilizers (as a source of phosphorous).

APROTIC SOLVENTS : Inert solvents which do not possess hydrogen, e.g., liquid sulphur dioxide, carbon tetrachloride, carbon disulphide, etc.

AQUA REGIA : A mixture of concentrated nitric acid and concentrated hydrochloric acid in the ratio of 1 : 3. It is a strong oxidising agent and can dissolve all metals, including noble metals. It is also called royal water. The mixture gives nascent chlorine and nitrosyl chloride (NOCl).

AQUEOUS : A solution in water.

ARARGONITE : $CaCO_3$. A type of calcium carbonate. It is less soluble and harder than calcite. With passage of time it undergoes recrystallistion to calcite. It occurs with limestone as a deposit in its cavities, as a precipitate around hot springs and in high pressure, low temperature metamorphic rocks. It is the main constituent of pearls. In pure state it is white or colourless but may look grey, blue, green or pink in the presence of impurities.

ARENES : Aromatic hydrocarbons like benzene, toluene, xylenes, napthalene, etc.

ARGENTIC OXIDE : *See silver (II) oxide.*

ARGENTITE : A sulphide ore of silver, Ag_2S. It is shining black in colour when freshly cut.

ARGENTOUS OXIDE : *See silver (I) oxide.*

ARGININE : It is an α- amino acid.

ARGON : Ar, A noble gas belonging to the 'O' group. It exists in air in a small amount in its non–atomic form. It's atomic number is 18 and relative atomic mass 39.948. It is obtained from liquid air by fractional distillation. It is colourless, odourless and only slightly soluble in water. It is used to maintain an inert atmosphere in welding, in the manufacture of metals like titanium and zirconium and in the filling of electric lamps. It was discovered by Lord Rayleigh and Sir William Ramsay.

ARNDT'S ALLOY : An alloy of 60% copper and 40% magnesium used for the estimation of nitrates.

ARNDT : Eistert synthesis - A process of converting a carboxylic acid into another having one carbon atom more or to an acid derivative.

AROMATIC COMPOUND : An organic compound which contains a benzene ring in its structure. These compounds do not show the chemical properties charateristic of unsaturated hydrocarbons (i.e. addition reactions) but exhibit substitution reactions. This unique property is due to the presence of delocalised π-electron, e.g., cloud over the ring. According to Huckel's theory, the delocalised π-electron cloud should be of $(4n + 2)$ electrons where n is a whole number integer.

The evidence in favour of delocalisation is (1) the bond lengths between carbon atoms in benzene are all equal – in between those of single and double bonds. (2) existence of only one ortho disubstituted derivative. (3) Benzene is more stable over its Kekule structure by 150 KJ mol^1

AROMATICITY : The characteristic property of extra stability of aromatic compounds; *see aromatic compounds*.

ARRHENIUS EQUATION : K = Ae $-E_a/RT$. An equation relating the specific reaction rate with activation. Ea and temperature A is the frequency factor – or pre exponential factor. Rate of reaction varies with temperature. The equation may be given in logarithmic form as :

$$\log K = \log A - \frac{E_a}{RT}$$

ARRHENHIUS THEORY : Theory of electrolytic dissociation of ionic substances in dissolved state. Strong electrolytes are almost completely ionised but weak electrolytes are much less ionised and maintain an equilibrium with the undissociated molecule. The extent of ionisation is called degree of dissociation.

On the basis of the extent of dissociation acids as well as bases may be classified as strong and weak acids and bases.

ARSENATE (III) : *See arsenic(III) oxide.*

ARSENATE (V) : *See arsenic (V) oxide.*

ARSENIC : Chemical symbol As; A metalloid element of group VA of the periodic table, having atomic number 33 and relative atomic mass 74.92. It exists in allotropic forms such as yellow, black and grey. The grey form is most stable. Arsenic is found in the natural state and also in several ores including mispickel (Fe As_5), realger (AS_5) and orpiment (AS_2S_2). The ore may be roasted in air to form oxide and then reduced by carbon to get arsenic. Various compounds of arsenic are used in insecticides, in medicines and as doping agents in semiconductors. It reacts with halogens, concentrated oxidising agents and hot alkalies.

ARSENIC (III) ACID : *See arsenic (III) oxide.*

ARSENIC (V) ACID : *See arsenic (V) oxide.*

ARSENIC (III) CHLORIDE : $AsCl_3$. A poisonous oily liquid. It is a covalent compound. In moist air it fumes because of hydrolysis.

$$2AsCl_3 + 3H_2O \rightarrow As_2O_3 + 6HCl$$

ARSENIC HYDRIDE : *See arsine.*

ARSENIC (III) OXIDE : As_2O_3 (arsenous oxide); A white or colourless, crystalline, poisonous solid. Also called white arsenic; exists as dimer As_4O_6. It is soluble in water, ethanol and alkali solutions. In nature it occurs as Arsenolite. Arsenic (III) oxide is obtained commercially as a by–product from the smelting of non–ferrous sulphide ore. It may also be produced in the laboratory by burning arsenic in air. Its structure resembles P_4O_6. Arsenous oxide is acidic and its solution is called arsenous acid. Its salts are called arsenate (III) salts or arsenites. It is used for the manufacture of opalescent glasses.

ARSENIC (V) OXIDE : As_2O_5; A white amorphous deliquescent solid. It is prepared by dissolving arsenic (III) oxide in hot nitric acid. It is soluble in water and ethanol. It is acidic and produces arsenic (V) acid (H_3AsO_4) formerly called arsenic acid. The acid is tribasic and gives arsenate (V) salts or arsenates.

ARSENIDES : Compounds of arsenic with metals.

ARSINE : Chemical fomulae AsH_3 arsenic hydride–a poisonous, colourless gas with an unpleasant peculiar smell. It is soluble in water, chloroform and benzene. It is produced by the reaction of mineral acids with arsenides of metals or by the reduction of many arsenic compounds using nascent hydrogen. Arsine decomposes on heating. It has pyramidal structure like ammonia or phosphine. It is used in a dilute gas mixture with an inert gas and its tendency of decomposition is used to enable other growing crystals to be doped with minute traces of arsenic to give n-type of semiconductors.

ARTINITE : A mineral of magnesium having composition Mg CO_3 $Mg(OH)_2$. 3 H_2O.

ARTIFICIAL RADIOACTIVITY : When a stable nucleus is bombarded with particles like neulras or α-rays, it produces a radioactive isotope of another element.

$$^{14}_{7}N \ + \ ^{4}_{2}N \ + \ ^{4}_{2}He \rightarrow \ ^{1}_{1}P + \ ^{17}_{8}O$$

Transuranic elements are all artificially radioactive as well as they are made by artificial radioactivity. None of them occur in nature. This is known as artifical radioactivity.

ARYL GROUP : A group obtained by removing a hydrogen atom from an aromatic hydrocarbon e.g. C_6H_5 phenyl group.

ASBESTINE : A type of talc mixed with tremolite asbestos, used in paints and as a rubber filler.

ASBESTOS : Any one of fibrous amphibole minerals like amosite, crocidolite, tremolite, anthophyllite, clirysolite. It is resistant to heat and electricity and inert to the attack of chemicals. It is used in making fireproof clothes, brake lining, table tops, etc.

ASCORBIC ACID : *See vitamin C*

ASPARGINE : $C_4H_8N_2O_3$. Aminosuccinamic acid. It is widely present in plants of lugirminosac and caraminae families and in several types of seeds, roots, etc.

ASPARTIC ACID : $C_4H_7NO_4$. Amino succinic acid. It is an alpha amino acid present in proteins.

ASPHALT : A natural mixture of bitumens.

ASPIRIN : Acetylsalicylic acid. An analgesic and antipyretic compound obtained by the acetylation of salicylic acid.

ASSOCIATION : The combination of molecules of a substance with those of another to form more complex species e.g. water and ethanol remain in the associated state due to intermolecular hydrogen bonding.

ASTATINE : Chemical Symbol At. Radioactive halogen (an element of group) VIIA having atomic number of 85 and relative atomic mass 210. It occurs in minute quantities with uranium and thorium ores. It can be made by alpha particle bombardment of bismuth. Twenty short lived isotopes of astatine are known, all of which are alpha particle emitters.

ASYMMETRY : Absence of symmetry in a molecule or a crystal.

ASYMMETRIC CARBON ATOM : A carbon atom linked to four different atoms or groups of atoms. Presence of such an atom in a compound causes optical activity.

ATOMOLYSIS : The method of separation of a mixture of gases based on their different diffusion rates.

ATMOSPHERE : (a) Earth's atmosphere (b) a unit of pressure equal to 101325 pascals or 760 mm/Hg.

ATOM : The smallest part of an element which may or may not exist independently but may participate in chemical reactions. It is made up of a nucleus (containing neutrons and protons) and orbitals containing electrons. The number of protons in all the atoms of an element remain constant and is called atomic number, while number of neutrons differ in different isotopes.

The electrons surrounding the nucleus are characterised by four quantum numbers which completely define size, shape and orientation of an orbital of the electron while the fourth quantum number represents only the direction of its spin.

ATOMIC ENERGY : Energy released when mass is converted into energy by the equation $E = mc^2$ where c is the velocity of light.

ATOMIC HEAT : It is the product of atomic weight and specific heat of the element. It is also called molar heat capacity of the element.

ATOMICITY : The number of atoms present in a molecule of element or a compound, e.g., Oxygen has atomicity 2, ozone has atomicity 3 and carbon dioxide 3.

ATOMIC MASS : *See atomic weight.*

ATOMIC MASS UNIT : (a.m.u) A unit of mass used for atoms and molecules to express their relative atomic masses. It is equal to 1/12 of the mass of an atom of carbon-12 ft is equal to 1.66×10^{-27}kg.

ATOMIC NUMBER : The number of protons and electrons present in the nucleus of an atom of element. It is the fundamental property of element. Position of element and all its properities is based on its atomic number.

Atomic Orbital : *See orbital.*

Atomic Volume : The ratio of relative atomic mass of an element to its density.

Atomic Weight : It is the weight of an atom of the element when compared to that of a C-12 atom taken as 12 a.m.u. It is called relative atomic mass.

Atropine : A poisonous crystalline compound. It is used in medicine to treat colic, to reduce secretions and to dilate the pupil of the eye.

Atto : A prefix used in S.I. system of units to denote 10^{-18} e.g. 1 attometre = 10^{-18}m.

Aufbau Principle : Principle governing the order of filling of orbits and orbitals of elements. The order is as follows: 1s, 2s, 2p, 3s, 3p, 4s, 3d, 4p, 5s, 4d, 5p, 6s, 4f, 5d, 6p, 7s, 5f, 6d.

This order is based on the energies of various orbitals. Different orbitals of an atom are filled in the order of their increasing energies. Energies of the orbitals may be compared by (n+l) rules which are as under (a) higher the summation (n+l), higher is the energy.

(b) if (n+l) values for two or more orbits are same, higher the value of n higher is the energy.

Auric Componds : Compounds of gold in its higher oxidation state, +3 e.g. auric chloride or gold (III) chloride, $AuCl_3$.

Aurous Compounds : Compounds of gold in its lower oxiation state, +1 e.g. aurous chloride or gold (I) chloride, AuCl.

Autocatalysis : A reaction in which one of the products promotes the rate of reaction further e.g. Mn^{++} ions act as catalyst in the reaction of $KMnO_4$ and oxalic acid. The rates of such reactions increase with time due to the increase in the amount of catalyst (product).

Autoclave : A strong steel container used to carry out chemical reactions at high temperature and under high pressure.

AUTUNITE : $Ca(UO_2)_2 (PO_4)_2nH_2O$, a mineral of uranium.

AUXINS : Plant hormones.

AUXOCHROME : A group in a dye molecule which influences its colour. These groups are -OH, $-NH_2$, etc. containing lone pair of electrons that may be delocalised along with the electrons of chromophore.

AVOGADRO'S LAW : The law states that equal volumes of all gases under similar conditions of temperature and pressure possess same number of molecules. The law is applicable to ideal gases. It was first proposed by Count Amadeo Avogadro.

AVOGADRO'S NUMBER : It is the number of particles (atoms, molecules, ions, electrons) in one mole of any substance. Its value is 6.02252×10^{23}.

AXIAL RATIO : The ratio of cell dimensions (a, b and c) in a particular crystal when b is taken as unity.

AZEOTROPE : (azeotropic mixture; constant boiling mixture). A mixture of two miscible liquids in a definite composition having constant boiling point i.e. the solution boiling as a pure liquid. The composition in the vapour phase and liquid phase remains same in such a solution. Azeotropes are formed due to the deviations from Raoult's law, leading to maximum or minimum in the boiling point-composition diagrams. The composition and boiling point of an azeotrope depends on temperature. An azeotropic mixture can not be separated by fractional distillation into its components.

AZIDE : Compounds containing azide ion N_3 (in inorganic compounds) or - N_3 group (in organic compounds).

AZIMUTHAL QUANTUM NUMBER : Angular momentum quantum number; Quantum number which denotes the shape of the orbital. Its values are equal to the value of n and range from 0 (zero) to (n-1) where n is the principal quantum number.

AZINE : An organic hetercyclic compound containing carbon and nitrogen atoms. Pyridine (C_5H_5N) is its simplest example. Diazines may have two, while triazines may have three nitrogen atoms.

AZOBENZENE : $C_6H_5=N - C_6H_5$ orange red coloured dye.

AZOCOMPOUNDS : Organic compounds containing $-N=N-$ group. They can be formed by coupling reaction of a diazonium salt with an aromatic phenol or amino compound. Such compounds are coloured and may come under the category of azo dyes.

AZODYES : An important category of compound used for dyeing wool and cotton. They are azo compounds and contain sone or more azo groups.

AZOIMIDE : *See hydrozoic acid.*

AZURITE : $Cu_3(OH)_2x\ (CO_3)_2$; basic copper carbonate; a mineral of copper formed in the upper zone of copper ore deposits. It is may be used as a blue pigment or as gemstone.

B

Babbit Metal : A group of related alloys used for making bearings. The alloy was originally made by a U.S. scientist, Issac Babbit. The alloy consists of tin containing about 10% antimony, 1-2% copper and a little lead.

Babo's Law : The law states that the vapour pressure of a solvent is lowered when a solute is added to it. The magnitude of lowering of vapour pressure is proportional to the amount of solute dissolved. The law was given by a German scientist, Lambert Babo.

Back Bonding : Overlapping of a filled orbital of an acceptor with a vacant d, p orbital of a ligand, forming sigma bonds.

Back E.M.F. : An electromotive force which opposes the normal flow of electric charge in a circuit. In an electric cell, back e.m.f. is set up due to the deposition of a layer of hydrogen bubbles on the cathode. Hydrogen ions take up electrons on the cathode and form gas molecules (polarisation).

Bacteriocidal : Substances capable of killing bacteria. Common examples are antibiotics, antiseptics and disinfectants.

Baddeleyite : An ore of Zirconium.

Bakelite : A trade name for certain resins made by condensation. Such resins were first introduced by the scientist Leo Hendrik Bakeland.

Ball Mill : A device used in the industry to decrease the size of solid substances. They have slowly rotating drums containing steel balls. The substance is crushed by the tumbling action of contents of the drum.

Baking Powder : Substances used in baking. These are generally the mixtures of sodium bicarbonate, potassium hydrogen tartarate or sodium hydrogen phophate together with some flour.

B.A.L. (BRITISH ANTI LEURISITE) : $C_3H_8OS_2$; an antidote to the poisoning by leurisite or other organic arsenicals as well as mercury, copper, zinc and cadmium.

BALANCE : A weighing device. Accurate laboratory balance can weigh up to fourth place of decimal of a gram. More modern substitutional balances have a single pan. These are more accurate and time saving. In automatic electronic balances mass is determined electronically. The mass of the empty container can be stored in the memory of the computer of the balance.

BALMER SERIES : A series of spectral lines obtained when the hydrogen atoms are excited by an electrical field. The lines correspond to the fall of excited electrons from higher energy levels to the second energy level. The radiation thus emitted fall in the visible region of the electromagnetic spectrum. The wavelength of the radiations of this series may be calculated by the equation :

$$v = \frac{1}{\chi} = R\left(\frac{1}{n_1^2} - \frac{1}{n_2^2}\right)$$

where x is wave number, is the wavelength, n_1 and n_2 are quantum numbers of which n_1 is 2 while n_2 may be 3, 4, 5– etc, R is Rydberg's constant (109678 cm^{-1}).

BANANA BOND : (a) Type of electron deficient bond (three centre-two electron bond) holding two boron atoms and a hydrogen atom as in boranes and similar compounds.

(b) In strained ring molecules, the angles calculated on the basis of hybridisation are generally not equal to the angles obtained geometrically by linking the atomic centres. For example, in cyclopropane the geometry suggests an angle of 60° while sp^3 hybridisation of carbon suggests it to be 109° 28^1. However, the actual angle is in between with the shape of a banana-like bent bond.

BAND SPECTRUM : A spectrum produced as a number of bands of absorbed or emitted radiations. It is made of groups or bands of closely arranged lines. These lines are the result of different vibrational states of molecules.

BAR : A unit of pressure which is equal to 10^5 pascals or 10^6 dynes per square centimetre. It is approximately equal to 750 mm Hg or 0.987 atmospheres. Millibar (10^{-3} bar or 10^2 pascals) is the commonly used unit of pressure in meteorology.

BARBITONE : $C_8H_{12}N_2O_3$. Derivative of barbituric acid; used as a sedative.

BARBITURIC ACID : $C_4H_4N_2O_3$; Malonyl urea.

BARFOED'S TEST : A test to identify reducing monosaccharide carbohydrates in solution. The test was first applied by a Swedish scientist, C.T. Barfoed. A reducing sugar like glucose gives a red precipitate with Barfoed's reagent (a mixture of acetic acid and copper (II) acetate).

BARIUM : Ba; An alkaline earth metal belonging to the second group of the periodic table. Its atomic number is 54 and relative atomic mass is 137.33. In nature, it occurs in the ores witherite ($BaCO_3$) and barytes ($BASO_4$). Barium may be extracted by reducing barium oxide with aluminium or silicon in vacuum or by the electrolysis of fused barium chloride.

The metal has a low ionisation potential, large size and thus strongly electropositive. It is readily oxidised in air and reacts with ethanol and water. Compounds of barium give a peculiar green colour on the flame of a burner. Soluble salts of barium are poisonous. The metal was discovered by Karl Scheele and was first extracted by Humphry Davy.

BARIUM BICARBONATE : $BaH(CO_3)_2$; Barium hydrogen carbonate; The compound is only stable in solution. It is formed by the action of carbon dioxide on barium carbonate suspended in cold water. $BaCO_3(S) + H_2O(l) \rightarrow Ba(HCO_3)_2$ (aq).

Barium bicarbonate on heating changes back to barium carbonate.

BARIUM CARBONATE : $BaCO_3$; A white crystalline compound, insoluble in water, decomposes on heating to barium oxide. It occurs in nature as the ore witherite. It may be prepared by adding an alkali metal carbonate to a barium salt solution. It is used as a rat poison.

BARIUM CHLORIDE : $BaCl_2$; A white solid which exists in two anhydrous, crystalline forms. The dihydrate of barium chloride $BaCl_2.2H_2O$ is prepared by dissolving the carbonate in hydrochloric acid and crystallising out the dihydrate. The dihydrate may lose water on heating to 386 K.

BARIUM HYDROXIDE : Ba $(OH)_2$; A white solid, slightly soluble in water, exists as octahydrate $Ba(OH)_2$. $8H_2O$. It is used in volumetric analysis to determine the strengths of weak acids using phenolphthalein as an indicator.

BARIUM OXIDE : BaO; A white solid substance obtained by heating barium in air or oxygen. It is also obtained by heating strongly, a sample of barium carbonate or heating in the presence of oxygen it forms the peroxide which may be used for the manufacture of oxygen.

BARIUM PEROXIDE :BaO_2; A white solid prepared by heating barium oxide in oxygen. It is used as a bleaching agent. With dilute acids, it forms hydrogen peroxide.

BARIUM SULPHATE : $BaSO_4$; A white solid which occurs in nature as barytes (or heavy spar). It is insoluble in water. It may be obtained by adding sulphuric acid to barium chloride solution. It is used in the glass and rubber industry under the trade name 'blanc fixe'. It is used as a pigment extender. It is opaque to X-rays and so its suspension in water is used as a medium for X-ray investigation of stomach and intestine.

BARIUM SULPHIDE : BaS; Black ash; highly phosphorescent.

BARN : b; A unit of area; 1b = 10^{-28} m^2. It is used to express the cross section of atoms and nuclei in the scattering experiments.

BARREL : A unit of volume used in industry; 1 barrel = 159 litres = 35 UK gallons.

BARYTA : $Ba(OH)_2$. *See barium hydroxide.*

BARYTES : (barites) Ba SO_4; heavy spar; a mineral form of barium sulphate.

BASE DISSOCIATION CONSTANT : K_b *See dissociation constant.*

BASE METAL : A common metal like iron, lead, copper, etc, which may corrode in air or moisture unlike the precious noble metals such as silver and gold.

BASE UNIT : A unit that is defined arbitrarily and not by the combination of other units. For example, ampere is a base unit in SI system while coulomb is the derived unit as the amount of change carried by one ampere current in one second.

BASIC : Denoting a compound which is a base or representing a solution containing larger concentration of OH ions. pH of basic solution is more than 7.

BASIC SALT : A compound in which one or more anions are replaced by oxide or hydroxide ions. Common examples include halides and carbonates. They are formed by the hydrolysis of metal salts and precipitated by sodium.

BASIC SLAG : *See slag.*

BASTENAESITE : It is cerium earth fluorocarbonate ($CeFCO_3$) containing traces of thorium.

BATCH PROCESS : An industrial process in which the reactants are added in fixed amounts called batches. Such a process lacks automation, and is only economical for preparing small quantities of valuable compounds like medicines, titanium metals, etc.

BATTERIES : A combination of several electric cells to produce current. These cells are connected in series. Total e.m.f. of the battery is the sum of e.m.f of all the cells joined together. The most common example is a lead accumulator consisting of six cells connected in a series to give a total e.m.f. of 12 volts. Such batteries are used in cars. If the cells are connected in parallel, the e.m.f of the battery remains same as that of one cell but their capacity is increased.

Capacity of a battery is described in ampere-hours (I ampere for 1 hour).

BAUXITE : Most important ore of aluminium consisting of hydrated aluminium oxide ($Al_2O_3.2H_2O$). It is an amorphous clay – like substance.

BECKMANN THERMOMTER : A thermometer used to measure small changes in temperatures. It is a mercury thermometer with a long, narrow stem with a scale covering 5 degrees calibrated in hundredth of a degree.

It is used to measure depression of freezing point or elevation of boiling point of liquids when solute is added. It can be set for any particular range.

It was invented by a German scientist, E.O. Beckmann.

BECQUREL – Bq : The unit of radioactivity. It is the activity of a radioactive nucleus decaying at a rate, on average of one spontaneous transition per second. 1 Curie = $3.7 x 10^{10}$ Bq. The unit is named after the discoverer of radioactivity, A. H Becquerel.

BEER'S LAW : A law relating the amount of light absorbed, the thickness of absorbing layer d and the molecular concentration*(c)*; $I = I_o e^{kcd}$ where I_o is the intensity of the incident radiation. I is the intensity of the transmitted radiation and k is molar absorption coefficient.

BEES WAX : Wax produced by bees and used in making shoe polishes and floor polishes. It is a mixture of mercyl palmitate. cerotic acid and its ester and some parafins.

BEILSTEIN'S TEST : A test for detecting the presence of halogens in an organic compound. When burnt on a copper wire a halogen containing organic compound gives a greenish colour to the flame.

BELL : Bell Metal - An alloy of copper (containing 60-85% copper) with tin and some zinc. It is used to make bells.

BENEDICT'S REAGENT : A reagent to test the presence of sugar. It is a mixture of copper (II) sulphate, sodium citrate and sodium carbonate.

BENEDICT'S TEST : A test to detect sugar and other reducing agents by Benedict's reagent. On heating the reagent with the sample solution, it gives a yellow precipitate (when reducing agent is in lower concentration) or red precipitate.

BENEFICATION : Also called ore dressing; the method of separating an ore from various impurities. It is achieved by several processes like crushing, grinding, froth floatation, magnetic separation, etc.

BENEFIELD PROCESS : A process for removing carbon dioxide gas from fuel gases.

BENT BOND : *See banana bond.*

BENZALDEHYDE : C_6H_5CHO;, It is a colourless liquid with almond – like smell (as it occurs in almond kernels). It is prepared by the controlled oxidation of toluene by Etard's method or by oxidising with CrO_3 and acetic anhydride. It is used in perfumery, flavouring and in the dyestuff industry.

BENZAMIDE : $C_6H_5CONH_2$; Colourless crystalline solid prepared by the ammonolysis of benzyl choride, benzoic anhydride and benzoic acid.

BENZENE : C_6H_6; Colourless aromatic hydrocarbon liquid with a characteristic odour. It was previously isolated as a component of coal tar distillation but now it is being made from the gasoline component of petroleum by catalytic reforming (catalyst platinium at 500°C and 10 atmospheric pressure). Benzene undergoes characteristic reactions of aromatic compounds.

Structure of benzene may be represented by resonance or by the delocalisation of π - electron cloud.

BENZENE CARBALDEHYDE : *See benzaldehyde.*

BENZENE CARBONYL CHLORIDE : C_6H_5COCl; benzoyl chloride. A liquid used as benzoylating agent.

BENZENE CARBOXYLIC ACID : *See benzoic acid.*

BENZENE : 1,4,–diol–$C_6H_4(OH)_2$; Hydroquinone; Quinol; A white crystalline solid used in making dyes.

BENZENE HEXACHLORIDE : $C_6H_6Cl_6$; Rindane; A crystalline solid made by the reaction of chlorine and benzene in the presence of light. It is used as a pesticide and insecticide.

BENZENE SULPHONIC ACID : $C_6H_5SO_3H$; A white crystalline solid made by the sulphonation of benzene with fuming sulphuric acid or chlorosulphonic acid. Its derivatives are used to make detergents.

BENZIDINE : $C_{12}H_{12}N_2$; Used as an indicator.

Benzoate Group : C_6H_5COO–group.

Benzoic Acid : C_6H_5 COOH; A white crystalline compound made by the oxidation of toluene by alkaline potassium permanganate. The sodium salt of benzoic acid is used as food preservative.

Benzoin : C_{14} $H_{22}O_2$; $C_6H_5CHOHCOC_6H_5$; It is prepared by the action of alcoholic sodium cyanide solution on benzaldehyde. The reaction is called benzoin condensation.

Benzophenone : $C_6H_5COC_6H_5$; Colourless crystalline solid, used in perfumery.

Benzoquinone : $C_6H_4O_2$; Cyclohexadine - 1,4 dione; Quinone; A yellow solid used in making dyes. It remains in equilibrium with hydroquinone in quinhydrone electrode systems; used to determine pH of solutions.

Benzoylation : A chemical reaction to introduce a benzoyl group into a molecule. It is similar to acylation and may be brought about by benzoyl chloride.

Benzoyl Chloride : *See benzene carbonylchoride.*

Benzoyl Group : C_6H_5CO-group.

Benzyl Alcohol : $C_6H_5CH_2OH$; phenylmethanol; A liquid aromatic alcohol; used as a solvent. It is synthesised by Cannizzaro's reaction, i.e., the simultaneous oxidation and reduction of benzaldehyde by aqueous caustic soda. The compound gives normal properties of alcohols as well as electrophilic substitutions of the benzene ring at position 2 or 4.

Benzyl Group : $C_6H_5CH_2O$–group.

Benzyne : C_6H_4; Aryne. A highly reactive intermediate which cannot be isolated. Benzyne and its derivatives occur as short-lived intermediates in organic reactions.

Bergius Process : A process used for making hydrocarbon mixtures (motor fuel like petrol) from coal, by heating powder coal, heavy oil and iron (III) oxide as catalyst in an atmosphere of hydrogen at about 723 K and 200 atmosphere. The process was developed by a German scientist, Friderich Bergius during World War-I.

BERKELIUM : A radioactive transuranic metal belonging to actinide series. Its atomic number is 97 and mass number of the most stable isotope is 24. It was discovered by G.T. Seaborg by bombarding a mercium 241 with alpha particles.

BERTHOLLIDE COMPOUND : *See non stoichiometric compound.*

BERYL : $3BeOAl_2O_3$. $6SiO_2$. A mineral form of beryllium aluminium silicate. It is the main ore of beryllium. It occurs in different colours like blue, green, yellow or white and is used as gemstones.

Emerald, the green variety due to traces of choromium oxide is of maximum value and occurs only rarely. Aquamarine, the blue-green variety, is another important form of beryl.

BERYLLATE : A compound formed in solution by the action of strong alkali on beryllium metal, its oxide or hydroxide. It may also be represented as $Be(OH)_4^{2-}$.

BERYLLIA : *See Beryllium Oxide.*

BERYLLIUM : Be. A grey alkaline earth metal; first member of II group of periodic table. Its atomic number is 4 while its relative atomic mass is 9.012. Minerals in which it is present are beryl ($3BeO.Al_2O_3.6SiO_2$); chrysoberyl ($BeO.Al_2O_3$); beryllonite ($NaBePO_4$); bertrandite ($4BeO.\ 2SiO_2$). It is extracted by the electrolysis of a mixture of BeF_2 and NaF or by reducing BeF_2 with Mg metal. The metal is used in making Be-Cu alloys used in nuclear reactors as moderators because of its low absorption tendency.

The metal is smallest in size, least electropositive, resistant to oxidation by air. Beryllium may react with oxygen, nitrogen, chlorine, sulphur at higher temperatures to form corresponding covalent compounds. It is amphoteric in nature and dissolves in dilute mineral acids as well as in caustic alkalies. It forms polymeric beryllium hydride $(BeH_2)_n$ by the reduction of $(CH_3)_2$ Be using $LiAlH_4$. In these hydrides two electrons and three centre bonds are involved, forming bridges between two Be atoms. It also forms coordination compounds e.g. $[Be\ Cl_4]^2$ $[Be\ (NH_3)_4]\ Cl_2$. The element is toxic and was isolated by F. Wohler and A.A. Bussy.

Beryllium Carbonate : $BeCO_3$; Solid and being less stable easily decomposes to beryllium oxide. It is formed by the reaction between beryllium hydroxide and carbon dioxide.

Beryllium Chloride : $BeCl_2$; A white solid obtained by passing chlorine over the mixture of beryllium oxide and charcoal. The anhydrous salt is used as a catalyst. It is a poor conductor even in the fused state indicating its covalent nature.

Beryllium Hydrogen Carbonate : $Be(HCO_3)_2$ Beryllium bicarbonate; It is formed in the solution state only when excess of carbon dioxide is passed through a suspension of beryllium carbonate. It is highly unstable and decomposes back to carbonate.

Beryllium Hydroxide : $Be(OH)_2$; It is obtained as a precipitate when a solution of caustic alkali is added to beryllium chloride (or some other salts). In excess of alkali, it dissolves to make beryllate.

Beryllium Oxide : BeO; Beryllia; A white solid, insoluble in water. It occurs in nature as bromellite and can be prepared by heating beryllium in oxygen or by thermal decomposition of beryllium hydroxide or by thermal decomposition of beryllium carbonate.

It is an amphoteric oxide. With acids it forms beryllium salts while with caustic alkalies it forms beryllates. It is used for making Be-Cu alloy which is used in refractories, high output transistors and integrated circuits. Its properties resemble those of aluminium oxide.

Beryllium Sulphate : $BeSO_4$; A white insoluble solid obtained by dissolving beryllium oxide in concentrated sulphuric acid. On heating, it decomposes to oxide. It is hydrolysed in hot water to give an acidic solution.

Bessemer Converter : A vertical cylindrical steel vessel lined with a refractory material. It is named after the British scientist, Henry Bessemer. It is used for making steel from cast iron or from copper pyrites.

Bessemer Process : Pig iron obtained from a blast furnace may be converted directly to steel in Bessemer converter (furnace) at about 1525K. Air is blown through molten pig iron and spiegel is added to introduce the correct amount of carbon. Impurities of silicon, phosporous, sulphur and manganese are removed as slag. The furnace may finally be tilted to get molten steel.

BETA DECAY: Radioactive decay in which beta particles or electrons are ejected from the unstable nucleus leading to the formation of a new nucleus. In this transfomation, neutron in converted into proton and electron and thus the atomic number is increased by one unit. For example, carbon-14 undergoes beta decay forming nitrogen-14.

BETA PARTICLES : *See Beta Decay.*

BICARBONATE : Hydrogen carbonate like sodium bicarbonate ($NaHCO_3$); made by replacing only one hydrogen of carbonic acid and so the formula for bicarbonate group or iron is $FeHCO_3$

BIMOLECULAR REACTION : A reaction involving the simultaneous collision of two particles (atoms, ions or molecules) e.g. the decomposition of hydrogen iodine to hydrogen and idione. All bimolecular reactions are second order reactions.

BINARY COMPOUNDS : A compound, alloy, acid formed from two elements e.g. NaCl, HI, etc.

BIOASSAY : An estimation test for substances having physiological effect.

BIOCHEMISTRY : The study of compounds and reactions occurring in the living organisms—plants and animals.

BIODEGRADABLE : Substances which may decompose naturally into simpler substances such as gases present in air and water. Such compounds are less harmful to our environment.

BIOSE: Carbohydrate having two carbon atoms, e.g., glycollic aldehyde-$CH_2(OH)CHO$.

BIOSYNTHESIS : The reaction by which the living organisms prepare various molecules needed for life, growth and reproduction.

BIOTIN : Component of vitamin B-complex. It acts as a coenzyme for various enzymes that catalyse the conversion of carbon dioxide into useful compounds. Its important sources are cereals, vegetable, milk, etc.

Biotite : A silicate mineral. It has sheet like crystal structure usually black or dark brown in colour.

Birkeland-Eyde Process : An industrial method for the preparation of nitric acid. The process involves the conversion of air to nitric oxide by passing through an electric arc. The process is preferred when cheap electricity is available. The method was developed by Kristian Birkeland and Samuel Eyde.

Bismark Brown : A basic azodye used as a hair dye.

Bismuth : A white (with a pink tinge) brittle metal belonging to group VA. Its atomic number is 83 while relative atomic mass is 208.98. It occurs in the native and combined state as well. Its ores are bismuthite, Bi_2S_3 and bismite Bi_2O_3. The metal is obtained from its oxide by reduction with carbon. It is used to make low melting alloys with tin and cadmium. It is also used as a catalyst for making acrylic fibres, and as a carrier of U-235. Bismuth compounds are used in cosmetics and medicines.

Bismuth (III) Carbonate Dioxide : Bi_2O_2-CO_3; Bismuthyl carbonate a white solid obtained as a precipitate when bismuth nitrate and ammonium carbonate solutions are mixed.

Bismuth (III) Chloride : $BiCl_3$; Bismuth trichloride; A white deliquescent solid made by heating bismuth in chlorine. In water it produces a white precipitate of bismuthyl chloride.

$BiCl_3 + H_2O \rightarrow BiOCl + 2HCl$.

This reaction is used to test for the presence of Bi^{3+} ions in a mixture.

Bismuth (III) Nitrate Oxide : $BiONO_3$; Bismuthyl nitrate; A white insoluble substance formed as a precipitate when bismuth (III) nitrate is diluted. It is also called bismuth subnitrate.

Bisulphate : – OSO_2OH group; Also called hydrogen sulphate.

Bisulphite : SO_3H group; Also called hydrogen sulphite.

Bittern : The solution of salts left after sodium chloride is crystallised out from seawater.

Bitumen : A mixture of solid or semi–solid hydrocarbons along with carbon obtained as a residue in petroleum refinery.

Bituminous Coal : Most abundant of various types of coal.

Bituminous Sand : A sandstone or porous rocky material containing hydrocarbons.

Biuret Test : A test to detect proteins in solutions. The given solution is mixed with a solution of sodium hydroxide and 1% copper sulphate solution is then slowly added. Formation of a violet ring indicates proteins or peptides.

Bivalent : Having a valency of two; also called divalent.

Black Damp : The air left after an explosion of methane in coal mines.

Black Lead : Another name of graphite.

Black Fixe : Barium sulphate when used as a pigment extender.

Blast Furnace : A tall, refractory lined cylindrical structure used for smelting of iron ores such as haemetite and magnetite to make iron. It is charged with concentrated ore, coke and limestone (which gives calcium oxide flux). Carbon monoxide produced within the furnace by the action of coke and hot air, reduces the oxides of iron. Silica present as impurity is removed on the form of slag which floats in the surface of molten pig iron.

Bleaching Powder : A white solid regarded as a mixture of calcium chloride, calcium and calcium hydroxide. It is prepared on a large scale by the action of chlorine and calcium hydroxide. It is used for bleaching paper, pulps and fabrics. It is also used for sterilizing water.

It is marketed on the basis of available chlorine. When treated with a diluted acid or even carbon dioxide, it liberates chlorine. Nascent oxygen thus produced by the action of carbon dioxide of air with aqueous solution of bleaching powder is responsible for its bleaching action.

Blende : A naturally occurring sulphide mineral like zinc blende, ZnS.

Block : A part of periodic table. Periodic table is divided into four blocks namely s,p, d and f block. *See Periodic Table for details.*

BLUE VITRIOL : $CuSO_4 . 5H_2O$; Copper (II) sulphate pentahydrate.

BOAT CONFORMATION : A typical shape of a molecule which arises by the rotation of its atoms or groups about C-C single bonds. The molecular structure looks like a boat and thus called boat conformation, e.g. boat form of cyclohexane.

BODY CENTRED CUBIC CRYSTAL (BCC) : A type of crystal in which the unit cell has an atom, ion or molecule at each corner of a cube as well as at its centre. In such a case coordination number is eight. Crystals of alkali metals have a bcc structure.

BOHEMITE : A mineral of aluminium AlO. (OH); It is named after a German scientist, J.Bohm.

BOHR THEORY : A theory introduced by Niel Bohr to explain the spectrum of hydrogen. He proposed that if an electron of mass m revolved in a circular orbit of radius r, with velocity v around a positively charged, central heavy particle nucleus, the total energy of electron (kinetic energy and potential energy) is determined as :

$$E = \frac{2\Pi^2 me_4}{n_2 h_2}$$

where e is the charge on an electron, n is the principal quantum number or orbit number and is Plank's constant. Further value of radius of orbits is also calculated as :

$$r = \frac{n_2 h_2}{4\Pi^2 me^2}$$

Angular momentum *mvr* of an electron may have fixed values like h / 2Π, 2h/2Π, 3h/ 2Π etc. It means that angular momentum is quantised and may be represented as Bohr assumed that when the atom emitted or absorbed radiation of frequency the electron jumped from higher to lower or lower to higher orbit. This theory could then explain spectral lines of hydrogen by the equation :

$$v = \frac{1}{\chi} = R \left(\frac{1}{n_1^2} - \frac{1}{n_2^2} \right)$$

Where is the wave number in the wavelength related with a spectral line and R is the Rydberg's constant whose value is $1.09678x10^7m^{-1}$ or $10978cm^{-1}$.

BOILING : The process of conversion of a liquid into its vapour state by heating at a constant temperature is called its boiling point.

BOILING POINT : The temperature at which the vapour pressure above the surface of the liquid equals the pressure of the surroundings. Boiling point of the liquid depends on the external pressure so its values are usually quoted for one atmospheric pressure.

BOILING POINT COMPOSITION DIAGRAM : A graph representing the relationship between the composition of the mixture and the boiling point (or vapour pressure).

BOLTZMANN CONTANT : K; $1,380622X10^{-23}Jk^{-1}$. It is equal to the ratio of gas constant (R) to the Avogadro's constant (N_o). It is named after an Austrian scientist, L. Boltzmann.

BOMB CALORIMETER : An apparatus used to experimentally determine the heats of combustion at constant volume (E). It consists of a steel calorimeter kept in a thermostat. In the steel calorimeter, a known weight of compound (food or fuel) is burnt in excess of oxygen by an electric spark. The heat of combustion is calculated from the rise in temperature involved.

BOND : The force of attraction which keeps atoms together in a molecule or molecules or ions in a crystal. The force may vary from very weak to very strong depending on the nature of bond (like ionic, covalent coordinate, metallic, hydrogen bond).

BOND ENERGY : An amount of energy released in the formation of a bond. It is usually expressed in terms of bond dissociation energy which is the amount of energy required to break a particular bond. For a compound having several bonds, average bond energies are calculated. The bond energies can be calculated from the standard enthalpy of formation of compound and from the enthalpies of atomisation of concerned elements.

Bonding Orbital : According to linear combination of atomic orbitals, two atomic orbitals on combination produce two molecular orbitals. The one of lower energy is called bonding orbital. It is responsible for the attraction between two atoms. It is more stable than the antibonding orbital.

Bone Ash : A white powder obtained by heating bones.

Bone Black : A residue obtained by the destructive distillation of bones. It is used as a bleaching agent in the sugar industry.

Boracic Acid : H_3BO_3; boric acid; ortho boric acid. A white crystalline solid, soluble in water. It is industrially prepared by treating kernite - a borate mineral ($Na_2B_4O_7$. $4H_2O$) with sulphuric acid. Its molecules exist as such in dilute solution but in more concentrated solutions, polymeric acids and ions are formed. It is a very weak acid. Solid boric acid on heating loses water forming another acid, $(HBO_2)_n$ called metaboric acid or polydioxoboric (III) acid. It is used in the manufacture of glass, enamels, glazes, paper etc. It is also used in medicines and as a food preservative.

Boranes : (Boron Hydrides). Compounds containing boron and hydrogen. They are of various compositions most common being diborane B_2H_6. Others include B_4H_{10}, C_5H_9 B_5H_{11}, B_6H_{10}. They are all volatile, reactive and are easily oxidised in air. The structure of boranes are quite different from conventional structures. Diborane for example has got four normal B-H bonds while two boron atoms are bridged through hydrogen atoms. Bonds thus formed are '2e–3c' (two electrons, three centre bonds) or banana bonds.

Borates : Compounds having a negative ion comprising boron and oxygen atoms. Lithium borate has anion $B(OH)_4$. Most of the borates are inorganic polymers having ring or chains based on Manar BO_3 group or tetrahedral $BO_3(OH)$ group. Hydrated boranes contain OH groups.

Borax : $Na_2B_4O_7$; Disodium tetraborate; A colourless or white solid soluble in water completely when heated to 593K. The compound contains $[BuO_5(OH_4)]^{-2}$ ions. The main sources are kernite, $Na_2B_4O_7.4H_2O$ and $Na_2B_4O_7.10H_2O$. It is an important substance in the glass and ceramics industries. In some cases, it is used as a flux also because such molten borates may dissolve in metal oxides. It is a salt of a weak acid and strong base and so is easily hydrolysed to weak alkaline solution. It is used as a mild antiseptic for the skin and mucous membranes. Borax is the source of compounds like barium borate (fungicidal paints), zinc borate (fire resistant substance) and boron phosphate (heterogenous catalyst in petrochemical industry).

BORAX BEAD TEST : A simple preliminary test in qualitative analysis for certain coloured metal ions. A transparent borax bead (or a membrane) is formed by heating a small amount of borax on to a loop of platinum wire. It is then allowed to touch the salt and the colour is observed in the oxidising as well as in the reducing flame. For example, cobalt gives a blue colour in hot as well as cold in both oxidising and reducing flames.

BORAZON : BN; Boron nitride; Solid, insoluble in water, decomposes in hot water. It is manufactured by heating boron oxide to 1073K in a current of nitrogen or ammonia in the presence of calcium phosphate. It is isoelectronic with carbon but is a non–conductor of electricity. It has a high thermal conductivity and high resistance.

BORIC ACID : *See Boracic Acid.*

BORIDE : A substance made by the direct combination of boron with metals (especially more electropositive) at high temperatures. They can also be made by the reduction of a mixture of the metal oxide and boron oxide with carbon or aluminium. The borides are all hard, high melting and conducting materials. Industrially, borides are used as refractory materials. Common examples include CrB, CrB_2, TiB_2, MgB_2, Zn B_2, etc.

BORN-HABER CYCLE : A thermodynamic cycle used for calculating the energies of formation of ionic crystalline solids. Various steps involved for M(s) are as under :

(a) Atomisation of the metal.

$M_{(s)} \rightarrow M_{(g)} : \Delta H_1$

(b) Atomisation of non-metal

$^1/_2X_2(g) \rightarrow X(g) : \Delta H_2$

(c) Ionisation of metal (ionisation energy)

$M_{(g)} \rightarrow M^+_{(g)} + e^-_{(g)}; H_3 \rightarrow X_{(g)} \Delta H_3$

(d) Ionisation of non-metal (electron affinity)

$X_{(g)} + e \rightarrow x(g) + e- \; x^-_{(9)} \; \Delta H_4$

(e) Formation of ionic solid (lattice energy)

$M^+_{(g)} + X^-(9) \rightarrow M^+X_{(s)} : \Delta H_5$

The resultant of all these is equal to enthalpy of formation of MX(s)

$H_f = H_1 + H_2 + H_3 + H_4 + H_5$ It is named after the German scientists, Max Born and Fritz Haber.

BORNITE : An important ore of copper Cu_5FeS_4 : also called peacock ore.

BORON : B; Hard, brittle, belongs to group IIIA of the periodic table. Its atomic number is 5 and relative atomic mass is 10.81. It occurs in nature as borax $Na_2B_4O_7.10H_2O$ and colemarite, $Ca_2B_6O_{11}$. It is obtained by the reduction of boron oxide by magnesium. In order to get highly pure boron for semiconductor applications, boron trichloride is reduced in vapour phase by hydrogen or electrically heated filaments. Due to its small size and high ionisation energy, it forms covalent compounds i.e. B^{3-} ion do not exist. Boron does not react directly with hydrogen. Boron burns in oxygen at a high temperature to form B_2O_3. It reacts with nitrogen at 1273K to form boron nitride.

Boron is used in semiconductors. Boron - 10 is used in nuclear reactors as control rods and shields.

Boron was discoverd by Humphry Davy, Gay-Lussac and L.J. Thenard.

BORON CARBIDE : B_4C ; A dark black coloured solid. It is extremely hard and is largely used as abrasive. It is also used as a neutron absorber in nuclear reactors.

Boron Hydrides : *See Boranes.*

Boron Oxide : B_2O_3; A white hygroscopic solid that forms boric acid. It forms various salts.

Boron Trichloride : BCl_3; A colourless fuming liquid which dissociates in water to give hydrochloric acid and boric acid. It is prepared by the action of dry chlorine with boron at high temperatures or by the action of chlorine on boron carbide at high temperatures. It is a Lewis acid due to electron deficient central boron atom. It may form stable addition compounds with ammonia and the amines. It is used for preparing various types of borones and for getting pure boron to be used in semiconductors.

Boron Trifluoride : BF_3 ; A colourless, fuming gas prepared by heating boron oxide, calcium fluoride and concentrated sulphuric acid.

Borosilicates : Compounds in which BO_3 and SiO_4 units are linked to form networks with a wide range of structures. Borosilicate glass is tougher, more heat resistant and has a wider plastic range, e.g. pyrex glass. Borosilicates are also used in glazes, enamels and making glass wool.

Bosch Process : It is an industrial process for preparing hydrogen gas from water gas by a heated catalyst. The hydrogen thus prepared is used for the manufacture of ammonia by Haber's process.

Boyle's Law : At constant temperature, the volume of a given mass of gas is inversely proportional to its pressure i.e. $P \alpha \frac{1}{V}$ or pV = constant. The value of the constant depends on the temperature and the nature of the gas. The law was discovered by an Irish scientist, Robert Boyle. It is also known as Mariotte's law after the scientist E. Mariotte.

Brackett Series : Series of spectral lines of hydrogen obtained when the electrons excited to higher orbitals jump down to the fourth orbit.

Brady's Reagent : A reagent (2,4 - dinitrophenylhydrazine) to test aldehydes and ketones. *See 2,4 dinitrophenyl hydrazine for details.*

BRAGG'S EQUATION : An equation used to study the structure of a crystalline substance by X-rays. $\lambda = 2d \sin Q$. where λ is the wavelength of the X-rays used, d is the distance between the planes and Q is the angle of incidence.

BRASS : A group of copper zinc alloys containing upto 50% zinc. Brasses containing upto 35% zinc can be worked cold and are specially suited for rolling into sheets, drawing into wire and making tubes. A typical yellow brass is of this type with about 33% zinc. Brasses having higher percentage of zinc are harder but less ductile.

BRAUNITE : Brown Mn_2O_3 mineral containing silica.

BRAZING METAL : Lead - tin alloy, a low melting alloy used in soldering.

BREUNERITE : Impure form of ore magnesite $MgCO_3$.

BRIDGING ATOM : An atom which joins two other atoms in a molecule e.g. hydrogen atoms in boranes.

BRINE : An aqueous solution of sodium chloride. It is used to extract sodium chloride or common salt. By its electrolysis, caustic soda is manufactured.

BRIM PROCESS : It was used to make oxygen gas by heating barium oxide in air which changes into barium peroxide. This peroxide on further strong heating produces oxygen.

BRITANNIA METAL : An alloy of tin (80-90 per cent), antimony (5-15 percent) and very small amounts of copper, lead and zinc. It is used in making bearings and several domestic articles.

BRITISH THERMAL UNIT : BTU : The unit of heat defined as the amount of heat required to raise the temperature of one pound of water by 1°F. 1BTU = 1055.06J.

BROMATES : A salt of bromic acid; salts possessing BrO_3^- ions.

BROMIC(I)ACID : HOBr; Hypobromous acid; A yellow liquid made by the action of mercury (II) oxide and bromine. It is a weak acid in aqueous solutions. It also acts as a bleaching agent as well as a strong oxidising agent.

BROMIC(V)ACID : $HBrO_3$; Bromic acid ; A colourless liquid prepared by the action of dilute sulphuric acid or barium bromate. It is strong acid.

BROMIDE : Salt of hydrobromic acid or salt containing Br ions.

BROMINATION : A reaction in which a bromine atom may be introduced into a molecule, e.g., bromination of benzene leads to the formation of bromobenzene. The reaction is carried out by liquid bromine in presence of suitable catalyst.

BROMINE : Br; An element of group VIIA (halogen family) of the periodic table. Its atomic number is 35 and average relative mass is 79.909. It is deep red in colour. At room temperature it exists as a volatile liquid. In small amounts, it is present in sea water, salt lakes etc. but is much less abundant. It may be obtained by passing chlorine gas through a bromide of mangenese dioxide and sulphuric acid or a bromide. Chemically its reactivity lies between chlorine and iodine. It forms a variety of compounds with oxidation numbers ranging from-1 to +7. Bromine is highly corrosive in its action and the vapours irritate the eyes and throat. Bromine is used in pharmaceuticals, photography, chemical synthesis and as fumigating agents. Large amounts of bromine is being used in the synthesis of 1,2-dibromoethane which is used along with anti knocking agent, tetraethyl lead. It was discovered by Antoine Balard.

BROMINE TRIFLUORIDE : BrF_3; Colourless fuming liquid. It is very reactive. It is made by the action of fluorine and bromine.

BROMOETHANE : C_2H_5Br; Ethyl bromide: A colourless volatile liquid used as a refrigerant. May be prepared by the reaction of hydrogen bromide with ethene.

BROMOFORM : $CHBr_3$; Tribromomethane. A colourless liquid having boiling point 151°C.

BROMOMETHANE : CH_3Br; Methyl bromide; A colourless volatile liquid used as a solvent like other alkyl halides.

BROMOTHYMOL BLUE : An acid-base indicator which gives blue colour in alkaline solution and yellow colour in acidic solution. Its pH range is 6-8.

BRONZE : Group of alloys of copper and tin. Containing varying amounts of other metals like lead, zinc, phosphorous, aluminium. These are generally harder, stronger and more corrosion resistant than brass. Gun metal contains zinc (2-4%) while phosphor bronze contains phosphorous (upto 1%). Aluminium bronze contains no tin but it has aluminium (up to 10%). Silicon bronze with 1-5% silicon has high corrosion resistance while beryllium bronze with 2% beryllium is very hard and strong.

BROWNIAN MOTION : The regular random motion of small microscopic particles in a liquid or gaseous medium, for example, movement of carbon or dust particles in air or movement of particles as dispersed phase in a colloidal solution. It was first observed by a botanist, Robert Brown, while studying the pollen grains. It is the result of bombardment of the particles by the continuously moving molecules of the fluid medium.

BRUCITE : A mineral of magnesium with composition $Mg(OH)_2$.

BTX : A mixture of aromatic hydrocarbons – benzene, toluene and xylenes. This increases the octave number of the petrol when added.

BUCHNER FUNNEL : A special type of funnel with a perforated circular base on which a circular vacuum paper can be placed. It is then attached to a filter pump for fast filteration. It is named after a German scientist, Eduard Buchner.

BUFFER : A solution which resists the change in pH when a small amount of an acid or base is added. Acidic buffers contain a weak acid and its salt with a strong base, e.g., acetic acid, and sodium acetate. Basic buffers have a weak base and a salt of weak base with strong acid, e.g., ammonia solution and ammoniun chloride. These are said to be the mixture of reserve acid and reserve base, e.g., in an acidic buffer, molecular weak acid acts as reserve acid while reserve base is the conjugate base ions of the same acid.

pH values of buffers are calculated by applying Henderson's equation for acidic buffer-$pH = pk_a + \log [salt]/[acid)$ while for basic buffer.

$$pH = 14 - \left(pk_b - \log\frac{[salt]}{[base]}\right)$$

Natural buffers occur in living organisms where the biochemical reactions are very sensitive to pH changes. Examples are carbonic acid with bicarbonate ions. Buffers are used in medicines (in the intravenous injections) in agriculture and in some industrial processes like dyeing, fermentation and in food processing industry.

Buffer Capacity : It is the number of moles of an acid or a base added to a litre of buffer solution so as to change the pH by unity.

Bumping : Violent boiling of a liquid caused by strong heating. It can be minimised by putting pieces of porous pot, broken china or glass beads.

Buna Rubbern : Synthetic rubber made by the copolymerisation of butadiene and styrene (Buna-s) or butadiene and acrylonitrile (Buna-n) catalysed by sodium or sodium ethoxide.

Bunsen Burner : A gas burner with a hole at the bottom for air. In it, gas (petrol gas or oil gas) mixed with air is burnt. When the air is in small proportions the flame is luminous and smoky. Properly adjusted burner gives a flame with a pale blue inner core of incompletely burnt gas. It is named after a German chemist, Robert Bunsen.

Bunsen Cell : A primary cell in which cathode is zinc immersed in dilute sulphuric acid while the anode is made of carbon plates dipped in nitric acid. The two are separated by a porous pot. The e.m.f of the cell is 1.9 volts.

Burette : An apparatus, made of long, narrow, graduated tube used for adding variable volumes of liquid in a controlled and measurable way. A burette has a stop cock and a small bore exit jet so that only a drop may be added at a time. Burettes are widely used in volumetric analysis.

Buta-1,3-Diene : $CH_2{=}CH\text{-}CH\text{-}CH_2$; A colourless, unsaturated hydrocarbon gas made by the catalytic dehydrogenation of butane (from natural gas). It is used in the manufacture of synthetic rubbers (Buna-s and Buna-n). The diene is conjugated and more stable than that of isolated dienes due to the delocalisation of π-electron cloud.

Butanal : CH-CH_2-CHO; Butyraldehyde; A colourless liquid aldehyde.

Butane : C_4H_{10}; A gaseous saturated hydrocarbon obtained from the gaseous fraction of crude oil. It can easily be liquified and the liquid thus obtained, filled in cylinders under pressure is used as a cooking fuel.

Butandioic Acid : $HOOC(CH)_2COOH$; Succunic acid; A crystalline, weak, dicarboxylic acid which occurs in some plants. It is formed during the fermentation of sugar.

Butonioc Acid : CH_3CH_2-CH_2-COOH; Butyric acid; A colourless liquid, weak aliphatic acid with a rancid odour. Its esters are used for flavouring and perfumery.

Butanol : C_4H_9OH; Butanol-1 (n-butyl Falcohol) and butanol-2 (*See butyl alcohol*) are colourless volatile liquids obtained from butane. Butanols are used as a solvents.

Butanone : CH_3-$COCH_2$-CH_3; Methyl ethyl ketone; A colourless volatile, water-soluble ketone. It is prepared by the oxidation of 2-butanol and is used as a solvent.

Butenedioic Acid : HOOC-CH=CH-COOH; Exists in the form of two geometrical isomers. The cis-form is called maleic acid while trans-form is called fumaric acid. Both are colourless, crystalline solids used in making synthetic resins. Trans form is more stable than that of cis form. Cis form can lose water on heating and forms maleic anhydride.

Butene : C_4H_8 : May be butene-1 or butene-2. Colourless, unsaturated, hydrocarbon gases with peculiar unpleasant odour.

Butter Of Antimony : $SbCl_3$; A white deliquescent solid.

Butyl Group : The organic, straight chain group; CH_3-CH_2CH_2-CH_2.

Butyl Rubber : A type of synthetic rubber obtained by the polymerisation of isobutylene and isoprene.

Butyraldehyde : *See butanal.*

BUTYRIC ACID : *See Butanoic Acid.*

BY–PRODUCT : A compound formed with the main product during a reaction. Some of the by-products are industrially very useful. For example, chlorine is a by-product of Castner-Kellner electrolytic process for the manufacture of caustic soda.

C

Cadmium : Cd; A soft, bluish metal of group IIB of the periodic table. Its atomic number is 48 while its relative atomic mass is 112.41. The name is derived from the ancient name for calamine($ZnCO_3$). It is found associated with zinc ores. Its ore is greenockite(CdS). Cadmium is used in solders, in nickel-cadmium batteries and in electroplating. Cadmium and its compounds are poisonous in nature. Its properties are similar to zinc but it has higher tendency to form complexes. In nuclear reactors it is used as neutron absorbers. It was discovered by Stromeyer.

Cadmium Cell : A type of primary galvanic cell used as a standard. It gives a constant e.m.f. of 1.0186 volts. *See Weston cell for details.*

Cadmium sulphide : A yellow solid, insoluble in water which occurs in nature as greenockite. It is used as a pigment in semiconductors and fluorescent substances.

Caesium : Cs; A Soft silvery white metal of group IA of the periodic table. Its atomic number is 55 while its relative atomic mass is 132.905. It occurs in several silicate minerals like pollucite $C_5AlSi_2O_6$. It occurs in small amounts in several minerals, one such source being carnallite. It is largest in size among alkali metals, has minimum ionisation energy and so is used in photo electric cells. It is also used in atomic clocks, as a catalyst and in ion-propulsion systems in spacecrafts. Its isotopes are radioactive e.g. caesium 137 is a gamma emitter.

Caesium Chloride : CsCl; Formed by the direct reaction of the metal or its carbonate with hydrochloric acid. Crystal structure consists of alternate layers of caesium ions and chloride ions with the centre of the lattic occupied by caesium ion and eight chloride ions on the corners. Over all, crystal structure is body centred cubic structure with 8.8 coordination number.

CAFFEINE : $C_8H_{10}N_4O_2$; An alkaloid present in tea and coffee. It acts as a stimulant and diuretic, used in cola drinks also.

COMPOUND CAGE: Clatherate compounds of noble gases (like Krypton) with some suitable organic compounds (like 8-hydroxy quionoline).

CALAMINE : $ZnCO_3$; An ore of zinc used in lotions for sunburns and sore skins.

CALCINATION : A process of strong heating of ore in a limited supply of air or oxygen. It is used for oxides, hydroxides and carbonate ores. Referbatory furnace is generally used for calcination.

CALCINITE : A mineral of composition $KHCO_3$.

CALCITE : A mineral form of calcium carbonate occuring in limestone, chalk and marble. It crystallises in the rhombohedral system. It has the property of double refraction.

CALCIUM : Ca; A soft, low melting, reactive metal of group IIA of the periodic table. It is an alkaline earth element. It is widely distributed in the earth's crust. Its important ores are marble ($CaCO_3$), gypsum ($CaSO_42H_2O$), flourspar (CaF_2), appatite (CaF_2, Ca_3 $(PO_4)_2)_5$ and limestone ($CaCO_3$).

Calcium is extracted by the electrolysis of fused calcium chloride. It has low ionisation energy due to its large size. It is highly electropositive and a very reactive metal. On heating with oxygen, nitrogen, sulphur and halogens it forms oxide, nitride, sulphide and halides respectively. Calcium also reacts with hydrogen to form hydride. With boron, arsenic, carbon and silicon, it forms boride, arsenide, carbide and silicide respectively. Its salts impart a characteristic brick red colour to the flame of bunsen burner.

Calcium is used as a reducing agent in the extraction of metals like thorium, zirconium and uranium. For living organisms, calcium is essential and is needed for normal growth and development.

CALCIUM ACETYLIDE : CsCl; Calcium carbide. A colourless solid, manufactured by heating calcium oxide with coke at a high temperature of 2273K in an electric furnace. When water is added, it liberates acetylene.

Calcium Bicarbonate: $Ca(HCO_3)_2$; Calcium hydrogen carbonate. A compound stable only in solution, formed when calcium carbonate in water suspension is dissolved by passing carbon dioxide gas. It is water soluble. On heating it decomposes to calcium carbonate, releasing carbon dioxide. So temporary hardness is removed by boiling the water. Scales deposited on the walls of the boiling containers are due to the formation of calcium carbonate.

Calcium Carbide : CaC_2; *See calcium acetylide.*

Calcium Carbonate : $CaCO_3$; A white solid sparingly soluble in water. It decomposes on heating to produce calcium oxide and carbon dioxide. It is present in limestone, calcite and avargonite. It is used in making lime, calcium acetylide, carbon dioxide and is the main raw material for Solvay process. It is used for making glass, mortar and cement, and is used as flux in the extraction of several metals.

Calcium Chloride : $CaCl_2$; A white deliquescent solid, soluble in water. It exists in the form of a monohydrate, dihydrate and hexahydrate. Large quantities are formed as byproduct of Solvay process. Anhydrous calcium chloride can be obtained when hydrated salt is heated in a current of hydrogen chloride. Anhydrous molten calcium chloride is used as an electrolyte for producing calcium. It is also used in mines and roads to reduce dust problem. It is used as a dehydrating agent in desiccators.

Calcium Cyanide : $CaCN_2$; A colourless solid, prepared by heating calcium carbide in a stream of nitrogen at about 1073K.

It is used as a fertilizer because with water it decomposes slowly to form ammonia. It is used in the manufacture of resin melamine, urea and some cyanide salts.

Calcium Dicarbide : CaC_2; *See calcium acetylide.*

Calcium Fluoride : CaF_2; A white crystalline solid. It occurs in nature as fluorspar (or fluorite). In the fluorite structure (of calcium, floride structure) each calcium ion is surrounded by eight flouride ions arranged at the corners of the cube while each fluoride ion is surrounded by four calcium ions arranged tetrahedrally.

CALCIUM HYDROGEN CARBONATE : $Ca(HCO_3)_2$. *See calcium bicarbonate.*

CALCIUM HYDROXIDE : $Ca(OH)_2$; Slaked lime. A white solid, sparingly soluble in water. It is made by the quenching of lime. The reaction is highly exothermic and is also called slaking. It is used to neutralise the acidic soil, in white wash, in the manufacture of mortar, bleaching power and glass. It may also be used to remove temporary hardness of water.

CALCIUM NITRATE : $Ca(NO_3)_2$; A deliquescent water, soluble in salt which crystallises as $Ca(NO_3)_2 .4H_2O$. On heating it first changes into its anhydrous state but on strong heating it decomposes to give calcium oxide, nitrogen peroxide and oxygen. It is sometimes used as a nitrogenous fertilizer.

CALCIUM OXIDE : CaO; Quick lime. A white solid formed by the thermal decomposition of limestone or calcium nitrate or by heating calcium in oxygen. Industrially it is manufactured by heating limestone. It is used as a flux in metallurgy, as a drying agent and to prepare calcium hydroxide.

CALCIUM PHOSPHATE : $Ca_3(PO)_2$; A white solid, insoluble in water. It is found in the minerals rock phosphate and apatite. It is an important component of animal bones and is used as a fertilizer.

CALCIUM SULPHATE : $CaSO_4$; Anhydrite; A white solid ; sparingly soluble in water. It occurs abundantly as anhydrite. Its more common form is gypsum ($CaSO_4.2H_2O$). Gypsum on heating under controlled conditions forms $CaSO_4. H_2O$ – plaster of Paris. This sets as a hard gelatinious mass with water. It is used in paints, in paper and ceramic industries.

CALGON : Trade name for a water softening agent. It consists of complex polyphosphate molecules which combine with dissolved calcium and magnesium ions to form hard water.

CALICHE : An impure commercial form of sodium nitrate. A typical sample may contain sodium, sodium chloride, sodium sulphate, calcium sulphate, magnesium sulphate, potassium nitrate, sodium metaborate, potassium chlorate and sodium iodate.

CALIFORNIUM : Cf: A transuranic, radioactive metal, silver grey in colour, obtained by the neutron bombardment of Americium 243 and Curium-244. Its atomic number is 98 while mass number of most stable isotope is 251. Californium-252 is an excellent neutron source and thus it is useful for radiotherapy. It was first prepared by G.T. Seaborg.

CALOMEL : Hg_2Cl_2; A white salt, insoluble in water and is prepared by heating mercury (II) chloride with mercury. It is used in calomel electrodes and cells and as a fungicide.

CALOMEL ELECTRODE : A half cell in which the electrode is mercury coated with calomel while the electrolyte is aqueous potassium chloride with saturated calomel. It acts as a standard electrode or half cell. Standard electrode potential is -0.2415 volts at 298K.

CALORIE : The amount of heat required to raise the temperature of 1g of water by $1^{\circ}C$. It is the unit of energy in c.g.s system. 1 calorie=4. 1868 joules.

CALORIFIC VALUE : Amount of heat liberated when a unit mass of food or fuel is completely burnt in excess of air or oxygen. The values are expressed in kilojoule per gram. Calorific values may be measured by a bomb calorimeter.

CALORIMETER : An apparatus for the measurement of thermal properties of substances like thermal capacity, calorific value, enthalpy changes during a chemical reaction etc.

CALX : A metal oxide formed when an air is heated strongly in air.

CAMPHOR : $C_{10}H_{16}O$; A white crystalline, cyclic ketonic compound. Originally obtained from Formosan camphor tree, it can now be synthesised. The compound has a peculiar smell. It is used in medicines as a stimulant. It is used in moth balls, ointments for colds and in the manufacture of explosives.

CANADA BALSAM : A resin used for mounting permanent slides for microscopic study. It is also used to join two quartz pieces to make a Nicol prism. Its properties are similar to that of glass.

CANDELA : Cd; It is the S.I. unit of luminosity or luminous intensity.

It is defined as the intensity in the perpendicular direction of the black-body radiation from an area of 1/600,000 signature metre at the temperature of freezing platinum under a pressure of 10,1325 pascals.

CANE SUGAR : $C_{12}H_{22}O_{11}$; Sucrose. *See sucrose for details.*

CANNIZZARO REACTION : A reaction of aldeydes, which do not possess any alpha hydrogen atom, to give corresponding alcohol and carboxylic acid salts in the presence of concentrated caustic alkali solution. The reaction involves the oxidation of one molecule and the reduction of another molecule simultaneously. For example, methanal is converted into sodium methanoate and methanol while benzaldehyde produces sodium benzoate and benzyl alcohol. It was discovered by the Italian scientist S. Cannizzaro.

CAPILLARY ACTION : The principle of rise of liquid in a capillary tube when it is dipped in the same.

CAPRIC ACID : n-C_9H_{19} COOH; *See decaoric acid.*

CAPROIC ACID : $CH_3(CH_2)_4$ COOH; A liquid carboxylic acid. Its glycarides occur in cow milk and in some vegetable oils.

CAPROLACNUM : $C_6H_{11}NO$; A white crystalline compound. It is a seven membered ring compound having -NHCO group and five -CH_2 group. It is used in the synthesis of nylon-6.

CAPRYLIC ACID : $CH_3(CH_2)COOH$; A colourless straight chain, saturated liquid, carboxylic acid.

CARAT : (a) A measure of quality of gold. Pure gold is assigned the value 24 carat. (b) A unit of mass equal to 200 milligram. It is used to measure the masses of diamonds, gems and other precious stones.

CARBAMIDE : Another name of urea. *See urea for details.*

CARBANION : A negatively charged organic ion which has the negative charge on a carbon atom. Carbonions are usually formed by abstracting a proton from a C-H bond with the help of a base. For example, carbanion-CH_2CHO is formed by the action of caustic alkali on ethanal during aldol condensation.

CARBENE : A species of the type RR'C : or R_2C : or simply : CH_2 in which a carbon atom has two non bonded electrons. Carbenes are highly reactive intermediates which exist in some reaction. They readily attack C=C double bonds to form cyclopropane derivatives.

CARBIDE : A compound of carbon with a metal or electropositive element. Some carbides are ionic in nature and have C^{4-} ions e.g., Al_4C_3. Such carbides react with water to form methane and so are also called methanides. Another type of carbide is of dicarbides which have C_2^{2-} ions, e.g., CaC_2. Such carbides produce acetylene when put in water and so are called acetylides. Both the above types of carbides are ionic crystalline solids and are formed with highly electropositive metals. Some carbides are covalent in character e.g., SiC, B_4C_3. Such carbides are hard and high melting solids. Transition metals form a range of interstitial carbides in which the carbon atoms occupy the interstitial positions in the metal lattice. These are hard, conducting solid materials e.g. TIC. Some transition metals are too small in size to accommodate carbon atoms. In such cases, the metal crystal lattice gets distorted and chains of carbon atoms may be accommodated as Cr_2C_2, Fe_3C. Properties of such carbides are intermediate between ionic and interstitial carbides. On reaction with water, they may give mixture of hydrocarbons.

CARBINOL : CH_3OH; methanol; Alcohols were previously named through the carbanian system, e.g., CH_3CH_2OH as methyl carbinol, $CH_3CH_2CH_2OH$ as ethylcarbinol etc.

CARBOCATION : R^+ - An intermediate cation in organic reactions in which a positive charge is accommodated on a carbon atom. It is formed by the heterolytic cleavage of a C-X or C-O bond of alkyl halides or alcohols etc.

Stability of carbocation is attributed to electron releasing effect of alkyl groups (+I effect) or to the resonance in case of aryl carbonium ions. The order of stability of some of these is given as - $3^0 >> 2^0 >> 1^0$.

CARBOCYCLIC COMPOUND : A ring compound which has a ring of carbon atoms or all the members of the ring are carbon atoms, e.g., benzene, cyclopentone etc.

CARBOHYDRATES : Polyhydroxy compounds having an aldehyde or a ketone group or compounds which when hydrolysed produce such compounds. These are the compounds which widely occur in nature in all living organisms. They may be of several types. Monosaccharides are simplest carbohydrates – generally sweet, crystalline and water soluble, e.g., glucose and fructose. Glucose is an aldose while fructose is a ketose. Polysaccharides, on the other hand, are more complex like starch cellulose, glycogen etc. These polysaccharides on hydrolysis produced disaccharides (sucrose, maltose, etc) and monosaccharides.

Carbohydrates perform a vital role in living organsims. Sugars like glucose and their derivatives are essential intermediates in the conversion of food to energy. Cellulose forms the cell walls in plants. Starch serves as energy stores in plants.

Structure of carbohydrates is quite complex. They have ring structure usually five membered (furnanose) or six membered (pyranose).

Carbolic Acid : An earlier name of phenol, C_6H_5OH.

Carbon : C. First element of group IVA of the periodic table. It is a non metal with atomic number 6 and relative atomic mass 12.001. It is the essential constitutent of all living matter. It occurs in free state as coal as well as in the combined state in the form of carbonates.

Carbon exists in the form of several allotropes. Two important allatropes are diamond and graphite. Diamond is an extremely hard, crystalline, covalent compound. In this, carbon atoms are linked to one another by covalent bonds making regular tetrahedral geometeries. C-C bond lengths are 154 picometres and bond angles are 109°28.'

Graphite is a soft, black, slippery substance also called black lead or plumbago. In graphite, the carbon atoms are arranged in layers in which each carbon atom is surrounded by three others. The layers are held together by much weaker Van der Waals forces of attraction. Graphite is a good conductor of heat and electricity. It is used for making electrodes, high temperature equipments, pencils and solid lubricant.

There are some amorphous allotropes of carbon also like carbon black, characoal etc. Some of the isotopes of carbon are radioactive (with mass numbers 10,11,14,15). Carbon-14 is used in carbon dating.

Chemically, carbon is unique in its property to make straight branched and ring compound by the self linking of carbon atoms. This property is called catenation.

Carbon-12 is considered as standard atom for measuring atomic masses. Mass of one twelveth of a C-12 atom is defined as one atomic mass unit.

CARBONATE-CO_3 : A salt of carbonic acid containing CO_3 - ion. Carbonate ion is triangular planar in structure. Metal carbonates, in general, are ionic in nature. Carbonates of alkali metals are water soluble and their aqueous solutions are alkaline in nature. Carbonates of other metals are insoluble.

CARBONATION : (a) Introduction of-COOH group into a molecule by the addition of carbon dioxide e.g., formation of carboxylic acid by the carbonation of Grignard reagents and preparation of salicylic acid by Kolbes reaction.

(b) Carbonation of soft drinks by adding carbon dioxide under high pressure.

CARBON BLACK : A type of carbon obtained by the incomplete burning (or partial combustion) of natural gas, petrol, kerosene etc. It is used as a black pigment for inks and paints. It is also used as a bleaching agent for oils and other organic liquid. In the rubber industry, it is used as filter.

CARBON CYCLE : It is the cycle through which carbon in the form of its compounds circulates in the environment. Carbon dioxide from the atmosphere is taken up by plants which convert it to carbohydrates by photosynthesis. It is transferred to animals which feed on vegetables and fruits. During respiration, plants and animals release carbon dioxide to the atomosphere. Combustion of fossil fuels like coal, petroleum and its derivatives also produce carbon dioxide. These fossil fuels are formed from dead animals and trees after decomposition.

CARBON DATING : A method of calculating the age of archeological materials of living origin. In living organisms, the ratio of carbon-14 to carbon-12 remains constant because plants are continuously absorbing carbon dioxide from the atmosphere (for photosynthesis) and emitting it back by respiration. In the atmosphere C-14 is being formed as a result of cosmic ray action.

When the organism dies or the tree is cut, C-14 intake stops but C-14 already present goes on emitting beta radiations, with a half life of 5730 years. Thus C-14 activity goes on decreasing. The ratio of C-12 to C-14 is thus a measure of the time passed since the death of the organism.

The method is valuable for specimens of upto 40,000 years old. For ages of upto approximately 7000 years the time scale has been calibrated by dendrochronology. It is based on the measurement of C-14 to C-12 ratio in tree rings whose age is known. The method was developed by W.F. Libby.

Carbon Dioxide : CO_2; A colourless, odourless, non flammable gas, soluble in water, ethanol and acetone. It is obtained by the burning of coal in excess of air or oxygen or by their thermal decomposition of limestone or by the action of mineral acids on metal carbonates. It occurs in the atmosphere in small proportions (0.03% by volume). Main use of carbon dioxide is in the formation of dry ice (solid carbon dioxide) requiring large scale refrigeration. It is used in fire extinguishers. It is also used in carbonated soft drinks.

It is suggested that excessive burning of fossil fuels will increase the overall amount of carbon dioxide. Consequently the temperature of the atmosphere will increase (green house effect).

Carbon Disulphide : CS_2; A colourless, highly refractive, poisonous, inflammable organic liquid. It is prepared by the action of methane (natural gas) and sulpher. It is almost immiscible in water but miscible in alcohol and ether. It is a very good solvent for oil, water, rubber, sulphur, etc.

Carbon Fibres : Fibres of carbon (graphite) made by heating stretched textile fibres and used to strengthen polyme.

CARBONIC ACID : H_2CO_3; A dibasic acid. It is an aqueous solution of carbon dioxide. It forms two types of salts namely bicarbonates and carbonates.

$H_2O + CO_2 + H_2CO_3 \Leftrightarrow [H^+ + HCO_3^-]$

$K_A - 4.5 \times 10^{-7}$ mol L^{-1}

$HCO_3^- \Leftrightarrow CO_3^{2-} + H^+$ $K_A = 4.8 \times 10^{-11}$ Mol L^{-1}

CARBONISATION : The process of conversion of an organic compound into carbon by incomplete oxidation on heating.

CARBON MONOXIDE : CO; A colourless, odourless, inflammable, poisonous gas, sparingly soluble in water but more soluble in alcohol and ether. It is formed by the incomplete combustion of carbon. It is an important constitutent of water gas. It is a very strong reducing agent and is used in several metallurgical processes. It can combine with several transition metals forming carbonyl complexes like $[Ni(CO)_4]$ and $[Fe(CO)_5]$. The property of nickel forming such a complex is utilized in Mond's process as the carbonyl complex formed decomposes back to metal on heating. Its tendency to form complexes makes it toxic in nature. It may form quite stable complex with haemoglobin where it forms a dative bond with iron and so blocks the property of haemoglobin as an oxygen carrier.

CARBON SUBOXIDE : (Tricarbon dioxide) C_3O_2; A colourless gas with an obnoxious odour. It is formed when malonic acid is strongly heated with phosphoric acid and so is called malonic anhydride.

CARBON TETRACHLORIDE : CCl_4; Tetrachloromethane – a colourless, volatile, non flammable liquid with a peculiar odour, almost immiscible in water but miscible in usual organic solvents. It is prepared by the chlorination of methane. It is a good solvent for oils, waxes, rubbers, etc. Its use is now being restricted as the moist carbon tetrachloride may slightly decompose to poisonous compound phosgene. It is used to manufacture freon. Under the name pyrene it is also used as fire extinguisher as its vapours are non inflammable.

CARBONYL GROUP : A complex in which carbon monoxide

molecules form coordinate bonds with transition metal atom. Examples are $Ni(Co)_4$ $Fe(Co)_5$, $Cr(Co)_6$ and $Mn_2(CO)_{10}$.

CARBONYL CHLORIDE : $COCl_2$; phosgene; A colourless, poisonous gas. It is formed by the aerial oxiation of chloroform or by the direct chlorination of carbon monoxide. Due to its highly toxic nature it was used during World War-I. It may be used as a chlorinating agent in organic reactions.

CARBONYL COMPOUNDS : Organic compounds having CO group. Aldehydes, ketones, carboxylic acids and carboxylic acid derivatives are common examples. Inorganic carbonyl complexes may also be included in this category. Nucleophilic addition, condensation polymerisation and oxidation reactions are some important chemical properties of carbonyl compounds.

CARBONYL GROUP : $C{=}O$ group present in aldehydes, ketones, acids and acid derivatives. Inorganic carbonyl complexes also have this group.

CARBORUNDUM : SiC; Silicon carbide. A black coloured, crystalline structure may be like zinc blende or wourtizite extremely hard, diamond like solid. It is prepared by heating silica with carbon in an electric furnace. It is widely used as an abrasive.

CARBOXYL GROUP : COOH group present in carboxylic acids. It is planar due to the presence of hybridised carbon atom. Due to the delocalisation of π-electron cloud it may liberate protons.

CARBOXYLIC ACID : Organic compounds containing-COOH, carboxyl group. They are weak acids. They may exist in liquid as well as in solid state. Depending on the number of carboxyl groups; they may be termed as mono-, di-and tricarboxylic acids. Long chain carboxylic acids occur in nature in the form of glycerides (in oils and fats) and so they are called fatty acids. They may be prepared in the laboratory by the oxidation of primary alcohols or aldehydes or by the hydrolysis of nitrites or cyanides by dilute mineral acids.

The carboxylic acids are more acidic than those of carbonic

acid and phenol. Their acidic properties are explained by the resonance stabilisation of corresponding carboxylate anion.

CARBURIZE : *See carbonisation.*

CARBYLAMINE TEST : It is a test to detect the presence of primary aliphatic or aromatic amino group. The test involves the heating of primary amine with chloroform and caustic alkali when an isocyanide with a peculiar, unpleasant and easily distinguishable smell is formed.

CARCINOGENIC : Cancer producing agents, e.g., tobacoo smoke, general industrial chemicals and radiations like X-rays and ultra violet rays.

CARIUS METHOD : A method for the quantitative estimation of halogens, sulphur and phosphorous in organic compounds. In this method an organic compound is strongly heated in a sealed pyrex glass tube with concentrated nitric acid so that the elements like halogens, sulphur and phosphorous change into their corresponding acids and may then be precipitated as silver halides. The precipitates thus formed may be filtered, washed, dried, and finally weighed to calculate the percentage of required element.

CARNALLITE : A mineral consisting of chlorides of potassium and magnesium. Its composition formula is $KCl\ MgCl_2.6H_2O$.

CARNOT CYCLE : The ideal efficient reversible cycle of four operations for a reversible heat engine. Four operations are as under :

(a) Isothermal expension at temperature T_1 with the absorption of q_1 heat.

(b) Adiabatic expansion with a fall of temperature from T_1 to T_2.

(c) Isothermal compression at temperature T_2 when q_2 heat is released.

(d) Adiabatic compression with a rise of temperature from T_2 to T_1. According to the theory proposed by a French scientist, N.L.S. Carnot, the efficiency of any reversible heat engine depends only on the temperatures of sources and sink (T_1 and T_2) and not on the properties of the working material.

Efficiency of the heat engine is given as the ratio of work done to the amount of heat absorbed during adiabatic expansion step.

CARNOTITE : $K_2(UO_2)_2(VO_4)_n.H_2O$; A radioactive mineral of uranium and vanadium with very a small amount of radium.

CARO'S ACID : H_2SO_5; Peroxosulphuric acid. A crystalline solid substance made by the action of hydrogen peroxide on concentrated sulphuric acid. It is a strong oxidising agent in aqueous solution.

CAROTENE : One of the compounds of carotenoids family of pigments. Common examples are B carotene and lycophene which impart colour to the carrot and ripe tomatoes. Carotenes release vitamin A when they get decomposed during digestion process.

CARRIER GAS : The gas used to carry the sample in gas chromatography.

CASCCADE LIQUIFIER : An apparatus for liquifying a gas having low critical temperature. To liquify one gas, another gas below its critical temperature is liquified and evaporated at low pressure to cool the first gas below its critical temperature. In actual practice, several steps are involved to completely liquify the given gas.

CASCADE PROCESS : A process taking place in large numbers of steps as one single step may not yield the proper desired result. In various uranium extraction processes, single step process gives only poor isolation and so the process needs to be repeated many times in order to achieve better results. Another example may be cascade liquifier.

CASE HARDENING : The process of hardening of the surface (of case) of steel used for the manufacture of glass, tools, crank shafts and other mechenical components. It is done by heating the steel in a hydrocarbon or by dipping the red hot metal into molten sodium cyanide. By this method of carburising, carbon content of the surface increases. Nitriding is another method of surface hardening which involves the diffusion of nitrogen into the surface layer to form nitrides. Both are sometimes employed.

Casein : A group of phosphoprotein found in milk. They are easily digested by the enzymes of young mammals and serve as a source of phosphorous.

Cassiterite : It is the principal ore of tin with the composition, tin (IV) oxide, SnO_2.

Cast Iron : A crude form of iron directly obtained from the blast furnace. It contains 2 to 5% carbon either as iron carbide (in white cast iron) or as graphite (grey cast iron). Sometimes phosphorous, sulphur and manganese may also be present. Cast iron is used on a large scale for making moulds. It may be converted to steel in a bessemer converter.

Castner-Kellner Cell : An electrolytic cell used in the manufacture of caustic soda. The cell consists of an outer compartment having graphite anode and brine as electrolyte while the inner compartment has iron cathode and dilute solution of caustic soda as electrolyte. The two compartments are separated by slate partitions dipped in a layer of mercury at the bottom. The mercury layer of outer compartment acts as cathode and that of the inner compartment acts as anode.

Catabolism : The reactions involving the breaking of complex organic molecules to simpler ones in living organisms. During catabolism, energy is released which is used for various biofunctions of the organism.

Catalysis : The phenomenon of change of rate of a reaction by the use of a substance known as a catalyst. *See Catalyst.*

Catalyst : A substance that alters the rate of a chemical reaction without itself undergoing any chemical change permanently. They are of two types such as positive and negative. Positive catalysts always increase the rate while negative catalysts (called inhibitors) decrease the rate. Catalysts having the same phase as the reactants are homogenous catalysts while those involving reactants in different phases are heterogenous catalysts. The catalyst provides an alternate path with lower activation energy. Position of the equilibrium remains unchanged. Some times a product of the reaction itself acts as a catalyst (autocatalysis).

Catalytic action is specific and requires some optimum conditions of temperature and pressure.

Catalytic Cracking : *See cracking.*

Catephoresis : *See electrophoresis.*

Catechol : 1,2–dihydroxybenzene. A colourless crystalline phenol used in developing photographs.

Catecholamine : A class of amines (like dopamine, adrenaline) which functions as neurotransmitters.

Catenation : The property by which atoms of an element join one another to form long straight, branched or closed chains. Carbon exhibits this property due to which a very large number of organic compounds are formed.

Cathetometer : A microscope or telescope fitted with cross wires in the eyepiece and mounted in such a way that it can move along a scale. Cathetometers are used to measure accurately very small distances.

Cathode : An electrode at which reduction takes place. In a cell cations change into neutral atoms by taking electrons from cathode. In a discharge tube, electrons are ejected from cathode and they move towards anode.

Cathodic Protection : It is a device by which the rusting of underground iron fitting is prevented. In this method, a highly electropositive metal like magnesium is connected to the iron article so that the iron article now acts as a cathode which is no longer corroded.

Cation : A positively charged ion, an atom or a group of atoms having some positive charge. Cations of atoms are formed by the loss of one or more electrons. The cations are attracted towards the cathode under the influence of some potential difference.

Cationic Detergents : *See detergent.*

Cationic resins : *See ion exchange.*

Caustic potash : KOH; Potassium hydroxide.

Caustic soda : NaOH; Sodium hydroxide.

Celestine : A mineral of Strontium-$SrSO_4$.

Cell : A device by which chemical energy is converted into electrical energy. A cell of this type – galvanic, voltaic or electrochemical cell–consists of two electrodes or half cells, two electrolytes separated from each other and a salt bridge or porous partition to link the two half cells. In a galvanic cell, the spontaneous reaction of half cells produce a potential difference between two electrodes. Example of such a cell may be Daniel cell; Cu-Ag cell. Electrolytic cell is another type of device in which the chemical decomposition is brought about at the cost of electrical energy. In such a cell, both the electrodes may be dipped in the same electrolyte, e.g., electrorefining of copper using copper sulphate as an electrolyte. Other types of cell may be dry cell, lead accumulator, concentration cells and fuel cells.

Cellophane : A transparent, highly inflammable sheet cellulose.

Celluloid : A transparent, highly inflammable, thermoplastic made from cellulose nitrate and camphor.

Cellulose : $(C_6H_{10}O_5)$; A polysaccharide that consists of long chain of glucose units. It is the main constituent of cell walls. Its fibrous nature is used in textile industry for the production of cotton and artificial silk.

Cellulose Acetate : A polymeric compound obtained by acetylating cellulose by treating it with acetic anhydride and acetic acid in presence of concentrated sulphuric acid. It is used in plastic, acetate film, acetate rayons, non shatterable glass, varnish, etc.

Cellulose Nitrate : Also called nitrocellulose or gun-cotton. It is highly inflammable and is obtained by the treatment of cellulose with concentrated nitric acid. It is an ester of nitric acid and has $-CONO_2$ group (and not $C-NO_2$ groups as earlier proposed due to the name nitrocellulose). It is used in explosives (gun-cotton) and celluloid.

Celcius Scale : A temperature scale based on two fixed points. Melting point of ice taken as 0°C and boiling point of water as 100°C. The scale between these two is then divided into hundred equal parts or degrees. The magnitude of one degree Celcius is equal to one Kelvin. This scale was earlier known as centigrade scale (up to 1948). Later on it was changed to celcius scale after the name of a Swedish scientist, Anders Celcius who in fact devised the inverted form of this scale (melting point of ice as 100°C and boling point of water as 0°C).

CEMENT : A substance used for binding loose materials together and setting to a hard mass. The common cement called Portland cement is a mixture of calcium silicates and aluminates. It is manufactured by heating limestone with clay. The substance thus formed is well pulverised. A little of gypsum is added to suitably increase the setting time. Setting of cement involves the formation of hydrated aluminates and silicates with water.

CEMENTITE : A constituent of some cast iron and steel to enhance the hardness.

CENTI : C; A prefix used in SI. System to denote 10^{-2} or one hundredth e.g., 1 centimetre $1cm = 1x10^{-2}$ metres or $10^{-2}m$.

CENTIGRADE SCALE : *See Celcius scale.*

CENTRIFUGAL PUMP : A mechanical device used for transporting liquids. *See pump also.*

CENTRIFUGE : A device for rotating a container at high speed used to hasten the sedimentation of suspensions. In this the particles of higher density move downwards and settle down easily. It may also be used to separate two immiscible liquids.

CERAMICS : It is the art of making useful articles from raw materials of an earthy nature by applying high temperatures. The most important raw material is clay aluminium silicate Al_2O_3 SiO_2. $2H_2O$, china clay or kaolin which is decomposed, feldspar (K $AlSO_3O_8$) mixed with little of quartz and mica is commonly used. Examples are pitcher jugs, glazed potteries, tiles etc.

CERIUM : Ce; A silvery metal of lanthanides series with atomic number 58 and relative atomic mass 140.12. It occurs in association with other lanthanides. Important minerals are monazite, allanite, bastnasite, and cerite. Its isotopes occur in nature and several isotopes have been found to be radioactive. Cerium is used in making the alloy 'mischmetal' containing 25% cerium for use in lighter flints. Ceria, the cerium oxide is used in glass industry. Cerium is used as a catalyst also. It was discovered by M.H.Klaproth.

CERMET : A material consisting of ceramic and a sintered metal. Used when high resistance to corrosion, temperature and abrasion is required.

CERUSSITE : An ore of lead, $PbCO_3$; Lead (II) carbonate. In pure form, it is white but may be grey due to the presence of impurities of metal sulphide.

CETANE : $CH_3(CH_2)_{14}\ CH_3$; Hexadecane; A saturated hydrocarbon component of diesel fuel.

CETANE NUMBER : It is the percentage by volume of cetane in the mixture of cetane and - methyl naphthalene which has same ignition quality as the fuel under examination when burnt in a standard engine separately.

C.F.C. : Chlorofluorocarbons, compounds used in refrigeration, airconditioning as aerosol propellents; cause ozone layer depletion.

C.G.S. SYSTEM OF UNITS : A system of units that utilises centimetre, gram and second as the units of length, mass and time. For all scientific purposes, it is being replaced by SI units.

CHAIN : A sequence of atoms of the same element in a molecule. It may be straight as well as branched. In a straight chain, an atom is not linked to more than two atoms while in a branched chain an atom may be linked to more than two atoms making side chains. A chain may be closed to form a ring when two ends of an open straight chain come closer and join.

CHAIN REACTIONS : A reaction in which one of the products contributes to the propagation of reaction through subsequent chain reactions usually involving free radicals as intermediates. An example of chain reaction may be halogenation of alkanes (e.g. chlorination of methane). In this, a chlorine molecule splits homolytically to make free radicals and the step is called chain initiation. The free radicals thus obtained are highly reactive, unstable species which attack alkane molecules forming new free radicals. These steps are called chain propagation. Free radicals sometimes join one another to form stable molecules and such steps are called chain termination. A reaction between hydrogen and chlorine forming hydrogen chloride in the presence of light is another common example of chain reactions. Nuclear fission reaction i.e., fission of U-235 by neutrons is example of chain reactions.

CHAIR CONFORMATION : *See conformations.*

CALCEDONY : A mineral consisting of microcrystalline variety of quartz. It occurs in large number of semiprecious gemstones like jasper, agate, tigers eye, etc.

CHALCOGENS : Ore producing elements; The elements of group VIA or oxygen family.

CHALCONIDES : Binary compounds formed between metals and chalcogens like oxides, sulphides, selenides and tellurides of metals.

CHALCOPYRITE : $CuFeS_2$; Copper pyrites; A yellow coloured ore of copper. It is the most widely available ore of copper.

CHALK : A natural form of calcium carbonate formed by sea organisms. Shells of eggs are also of calcium carbonate. It is used in tooth pastes.

CHARCOAL : A porous, amorphous form of carbon made by the destructive distillation of organic matter or by heating wood in the absence of air. All forms of charcoal are used as absorbents and purifiers for liquids. Animal charcoal or bone charcoal is made by heating bones and removing phosphates with the help of acids. It is used in sugar refining. Activated charcoal is charcoal activated by steam or by heating in a vacuum.

CHARGE : A property of some elementary particles generated when an atom gains or loses electrons. It may be positive or negative two particles having similar charges which repel each other. The unit of charge is coulomb. The charged carried by an electron is 1.602×10^{-19} coulomb commonly called 1 unit charge.

CHARLE'S LAW : The volume of a given mass of an ideal gas increases or decreases by 1/273 of its volume at 0°C for every 1°C rise or fall of temperature, at constant pressure. The law leads to the fact that at -273°C, volume of gas must become zero which is not possible according to the law of conservation of mass. Such a temperature can never be attained in a gas phase and so is called absolute zero or zero kelvin. A scale of temperature starting from 0K is called absolute or Kelvin scale. The law may also be given in a simpler form which states that volume of a given mass of gas is directly proportional to its temperature on Kelvin scale, keeping the pressure constant.

$V = kT$ at constant pressure.

A similar relationship can be given for pressure also keeping the volume constant.

The law was given by a French scientist, J.A.C. Charles, but was independently and more accurately established by the experiments of another French scientist, J. Gay-Lussac, and so sometimes it is also called Gay-Lussac's law.

CHEDDITE : A group of high explosives made by mixing nitro compounds with potassium chlorate.

CHELATE : An inorganic coordination compound having a ring of atoms including the metal. Ligands forming chelates are called chelating agents and they are bidentate (ethylene diamine, oxalato, etc) or polydentate (ethylene diamine tetra acetato legands). The process of its formation is called chelation. The word is taken from the greek word meaning claw.

CHEMICAL BOND : A force of attraction between atoms holding them together in a molecule or a crystal. The atoms may be of the same element or of different elements. The chemical bonds may be of several types such as covalent, ionic, coordinate.

CHEMICAL COMBINATION : Compounds are formed by the chemical combination of elements. Qualitative and quantitative study of chemical reactions led to the discovery of various laws of chemical combination. They are – law of conservation, law of multiple proportions and law of equivalent proportions (separately dealt at appropriate places).

CHEMICAL DATING : A dating technique based on measuring the composition of a given sample. It is used when the sample undergoes slow chemical change at a known rate for example phosphate in the dead bones is slowly replaced by fluoride ions from the ground water and measurement of the proportion of fluorine present gives an idea of the time that the bones have been there in the ground. Chemical dating may also be based on the measurement of relative amounts of d- and l - amino acids present in the bones because after death, amino acids of living organisms slowly racemize.

CHEMICAL ENGINEERING : The branch of engineering which deals with the design, manufacture, maintenance and operation of a chemical plant and other machinery in the industrial chemical processes. It enables laboratory processes to be modified into large scale commercial production of chemicals. It involves the study of suitability of conditions like temperature, pressure, catalysts, methods of separation and purification, storage with minimum corrosion and wear, etc.

CHEMICAL EQUATION : A brief representation of chemical rections with the help of symbols and formula of various reactants and products involved. An equation must be balanced with respect to all the atoms and molecules.

$$CaCO_3 + 2HCl \rightarrow CaCl_2 + H_2O + CO_2.$$

To overcome several limitations, an equation is now modified to represent physical states, concentration of the reactants, gases evolved or solids precipitated. Conditions can also be shown on the arrow separation reactants from products. Further, if the reaction is reversible, double arrow is used in between reactants and products. Equation described above may be given as -

$$CaCO_3(s) + \text{dil } 2HCl\ (ar) \rightarrow CaCl_{2(}ar) + H_2O(e) + CO_2(g)$$

CHEMICAL EQUILIBRIUM : A reversible chemical reaction, which, after sometime attains a state called equilibrium when the concentration of reactants and products do not change. It is due to the fact that rate of forward reaction equals to the rate of backward reaction, e.g.

$$Fe(s) + 4H_4O(g) \rightarrow Fe_3O_{4(}s) + 4H_2(g)$$

CHEMICAL EQUIVALENT : It is the weight of an element which may combine with or displace, directly or indirectly 1.008 parts by weight of hydrogen or 8 parts by weight of oxygen or 35.5 parts by weight of chlorine in a chemical reaction.

For an element, Equivalent weight = Atomic weight/valency.

For a salt, Eq. wt = Molecular weight/total positive valency of metal.

For an oxident Eqwt = Molecular weight / No. of electrons lost per molecule.

For an acid, Eq. wt = Molecular weight/basicity.

For a base, Eq. wt = Molecular weight/acidity.

CHEMICAL FOSSIL : Organic compounds found below the surface of earth which appear to be biological in origin and indicate that life existed when the rocks were formed.

CHEMICAL POTENTIAL : It is the change in free energy with respect to change in the amount of one component with pressure, temperature, the amount of other components being constant.

$\mu = G/n$

CHEMICAL REACTION : A change in which one or more elements or compounds react among themselves to form one or more new compounds. Usually bonds present in the reacting substances break and new bonds of products are formed. Most of the reactants are reversible when no product is allowed to escape out or settle down. During chemical reactions some energy is either absorbed or released, the amount of which depends on the bond energies of reagents involved.

CHEMICAL WARFARE : Chemical substances which are used in war. They may be toxic, inflammable and explosives.

CHEMILUMINESCENCE : The emission of light during a chemical reaction without rise in temperature. e.g., slow oxidation of phosphorous; light emitted by glow-worm is supposed to be due to the oxidation of a protein, luciferin.

CHEMISORPTION : A type of adsorption in which actual chemical bonds are formed between adsorbate molecule and adsorbent surface. It is highly specific, irreversible, single layered and possesses high bond energy.

CHEMISTRY : The branch of science which deals with the study of extraction and properties of elements as well as the methods of preparation and properties of compounds. It may simply be described as the study of matter. *For details see the branches like biochemistry, inorganic, organic, physical, analytical and geochemistry.*

CHEMOTHERAPY : A type of chemical treatment based on the destruction of pathogenic organisms.

CHERT : *See flint.*

CHILE SALTPETRE : A commercial mineral which contains sodium nitrate in large proportions. Its deposits are found in Chile. It is principally used as a fertilizer and as a source of nitrogen.

CHINA CLAY : Kaolin; A white powder used in the paper and pottery industry.

CHINESE WHITE : ZnO; Zinc oxide; Used as a white pigment and a wild antiseptic in zinc ointments.

CHIRAL CARBON : A carbon atom in a molecule which is attached to four different atoms or groups of atoms; also called asymmetric carbon atom. Pressure of a chiral carbon atom causes optical activity in the organic compound.

CHIRALITY : Molecular dissymmetry of the molecule or the property of molecule by which it exhibits enantiomerism.

CHLORAL : Cl_3CCHO; trichloroethanal. It is an intermediate in the manufacture of chloroform. It is used to make DDT.

CHLORAL HYDRATE : $Cl_3CH(OH)_2$; 2,2,2-trichloroethandiol.

CHLORAMINE : NH_2Cl ; A colourless liquid formed by the action of ammonia and sodium hypochlorite. It is an unstable compound that explosively decomposes to ammonium chloride and nitrogen trichloride.

CHLORATE : Salt of chloric acid, $HClO_3$, e.g., sodium chlorate, $NaClO_3$.

Chloretone : A hypnotic compound used in sleep inducing pills made by the condensation of chloroform and acetone.

Chloric (I) Acid : HOCl; Hypochlorous acid. A colourless liquid which is stable only in solutions. It is prepared by passing chlorine gas in water or by passing chlorine through a suspension of mercury (II) oxide. It is a very weak acid and a mild oxidising agent which is used as a bleaching agent also. At lower temperatures, in aqueous solutions, it disproportionates to make chloride and chlorate (V) ions.

Chloric (III) Acid : $HClO_2$; Chlorous acid. A light yellow acid which exists only in solutions. It is prepared by the action of chlorine dioxide and water. It is a weak acid and an oxidising agent.

Chloric (V) Acid : $HClO_3$; Chloric acid. A colourless, unstable liquid with pungent odour. It is prepared by the reaction of barium chlorate with dilute sulphuric acid. It is a strong acid and strong oxidising agent. In concentrated solution, it ignites organic substances like paper and sugar.

Chloric (VII) Acid : $HClO_4$; Perchloric acid. A colourless, unstable liquid which fumes in moist air. It is prepared by vacuum distillation of a mixture of potassium perchlorate and concentrated sulphuric acid. It is a very strong acid and a strong oxidising agent. It is used to decompose organic matter.

Chloride : Cl^-; Anion formed when a chlorine atom accepts an electron.

Chlorination : (a) A reaction in which a chlorine atom is introduced into a compound e.g., chlorination of benzene to get cholorobenzene.

(b) The treatment of water with chlorine to disinfect it and to make it fit for drinking.

Chlorine : Cl; A halogen element; second member of group VII A of the periodic table having atomic number 17 and relative atomic mass 35.453. It is a poisonous, greenish gas which occurs widely in nature as sodium chloride in sea water, salt lakes and underground deposits of halite, NaCl. It also occurs in many other ores e.g., carnallite and sylvine. Chlorine is produced as a by-product in the manufacture of caustic soda and sodium. Chlorine is a highly reactive non-metal and may directly react with a large number of metals to form chlorides. It may react with a large number of organic compounds to chlorinate them. It also behaves as a strong oxidising and bleaching agent. The compounds of chlorine contain chlorine in the various oxidation states like - 1, 0, 1, +3, +5 and +7. Chlorine is used for the chlorination of drinking water, manufacture of large number of inorganic and organic chlorocompounds and as a bleaching agent.

It was discovered by Karl Scheele.

Chlorine Dioxide : ClO_2; A yellowish red, unstable explosive gas which is soluble in cold water. It is formed by the action of concentrated sulphuric acid or potassium chlorate or by the action of sodium chlorate and moist oxalic acid at about 373K.

An aqueous solution of this gas is prepared on a large scale by passing nitrogen dioxide through a fused mixture of aluminium oxide and clay which is a solution of sodium chlorate. It is widely used as a bleaching agent in flour mills and in paper mills. It may also be applied to purify drinking waters.

Chlorine Monoxide : Cl_2O; An orange coloured gas prepared by the oxidation of chorine using mercury (II) oxide. It is also called chloric (I) anhydride.

Chlorite : (a) A chlorate (III) salt or a salt of chloric (III) acid (chlorous acid).

(b) A group of green or white layered silicate minerals. They are composed of silicates of aluminium, magnesium and iron with the formula $(Mg\ AlFe)_{12}\ (Si\ Al)_8\ O_{20}\ (OH)_{16}$. Chlorities are most common in igneous and metamorphic rocks.

Chloroacetic Acid : $ClCH_2COOH$; *See chlorethanoic acid.*

Chlorobenzene : C_6H_5Cl; A colourless, highly inflammabble liquid prepared by the direct chlorination of benzene using iron or ferric chloride as a halogen carrier. It is used as an industrial solvent as well as a starting material to prepare several organic compounds like phenol, aniline, benzonitrile DDT, etc.

2-CHLOROBUTA-1-3-DIENE :

$$\begin{array}{c} \quad\ \ Cl \\ \quad\ \ | \\ C = C - CH = CH_2 \end{array}$$

A colourless liquid used to manufacture synthetic rubber, neoprene.

CHLOROETHANE : CH_3CH_2Cl; ethyl chloride. A colourless, inflammable gas prepared by the action of hydrogen chloride on ethane. It is used as refrigerant, local anaesthetic and in making tetraethyl lead, an antiknocking agent.

CHLOROETHANOIC ACID : (Chloroacetic acids). Three chloroethanoic acids obtained by the replacement of hydrogen atoms of ehanoic acid one by one by chlorine atoms. These acids are termed as monochloro, dichloro and trichloroethanoic acids. Chlorogroups are electron withdrawing in nature and consequently chloro acids are stronger than the parent carboxylic acids.

These acids are prepared by the action of chlorine in presence of red phosphorous (Hell - Volhard - Zelinsky reaction).

CHLOROETHENE : $CH_2=CH\text{-}Cl$; Vinyl chloride. A gaseous compound formed by the action of hydrogen chloride or ethyne or by the action of chlorine on ethene and subsequent adehydrohalogenation by calculated amount of alcoholic caustic potash.

$$CH_2=CH_2+Cl_2Cl\text{-}CH_2\text{-}CH_2\text{-}Cl$$

$$Cl\text{-}CH_2CH_2\text{-}Cl+KOH\ (alcohol) \rightarrow CH_2=CH\text{-}Cl.$$

It is used to manufacture polyvinyl chloride, commonly known as PVC, an important polymer.

CHLOROFORM : $CHCl_3$ *See trichloromethane.*

CHLOROFLUOROCARBONS : CFC A group of compounds formed when some hydrogen atoms of hydrocarbon are replaced by chlorine and some by fluorine.

They are in general unreactive compounds used as refrigerants and in making plastic foams. Its use is now being restricted due to the feeling that in atmosphere such compounds break down and so damage the ozone layer.

CHLOROMETHANE : CH_3Cl; Methyl chloride. A colourless, inflammable gas obtained by the direct chlorination of methane (industrial method) or by the action of phosphorous (III) chloride or phosphorous (V) chloride on methanol. It is used as a local anaesthetic and as a refrigerant. It is converted to methyl magnesium chloride (Grignard reagent) which is an important synthetic tool.

CHLOROPHYLL : The green pigment present in plants which acts as catalysts in the photosynthesis of carbohydrates. Chlorophyll acts as a photosensitiser for this reaction. It absorbs energy of photon and then transfers it to the reactant molecules (carbon dioxide and water) which are unable to absorb radiation directly.

CHLOROPLATINIC ACID : H_2PtCl_6; Platinic chloride. A reddish crystalline substance prepared by the action of platinum with aquaregia. It is obtained by dissolving platinum (IV) chloride in concentrated hydrochloric acid. It is used to determine the equivalent weights and molecular masses of organic bases.

CHLOROPRENE : Chlorobuta-1,3-diene. A colourless liquid used in the manufacture of synthetic rubber, neoprene.

CHLOROSULPHANE : *See disulphur dichloride.*

CHLOROUS ACID : $HClO_2$; Chloric (III) acid.

CHOKE DAMP : *See blackdamp.*

CHOLOSTERIC CRYSTAL : *See liquid crystal.*

CHOLESTEROL : A harmful substance deposited on the walls of the arteries and the chief constituent of gallstones; it is a kind of alcohol called sterols. Sterol in turn belong to the category of steroids. It occurs in animal tissues and may be obtained step by step from isoprene unit.

CHOLINE : $CH_2(OH)$-CH_2-$N(CH_3)(OH)$; Occurs widely in living organisms as a constituent of phospholipids like lecithins and sphingomyelins.

CHROMATE : CrO_4 –; A salt containing chromate (VI) anions.

CHROMATOGRAM : A graphical record showing relative amounts of constituents in a substance developed during thin layer, paper or gas chromatography.

CHROMATOGRAPHY : A method used to separate mixtures of gases, liquids and dissolved solids by the application of the principle of absorption. The technique was developed by Russian scientist, Mikhail Tsvet, in the form of column chromatography. He used it to separate plant pigments which are separated into coloured bands by the elution (i.e. washing with a solvent). Chromatography involves two phases – the stationary phase (the adsorbent matter in the column like silica gel or alumina) and the moving phase or eluent. The separation depends on the adsorption of molecules of sample between the mobile phase and the stationary phase. Due to different extent of adsorption, substances get separated from one another. The substance which is least absorbed is collected first as eluate. In partition chromatography, a liquid (like water) is first absorbed by the stationary phase. Eluent is immiscible with it and separation is then achieved by means of partition between these two liquids.

CHROME ALUM : $K_2SO_4Cr_2(SO_4)_3$. $24H_2O$; Potassium chromium sulphate.

CHROME IRON ORE : A mixed ore FeO. Cr_2O_3, called chromite ore used to make chromium steels.

CHROME RED : PbO. $PbCrO_4$; Basic lead chromate used as a red pigment.

CHROME YELLOW : H_2CrO_2; A hypothetical acid.

CHROMIC OXIDE : *See chromium (III) oxide.*

CHROMIC COMPOUNDS : Compounds of chromium with oxidation number of +3 and +6.

CHROMITE : $FeOCr_2O_3$; The chief ore of chromium, black in colour with a metallic lustre.

CHROMIUM : Cr; A hard silvery transition element of 3d series having atomic number 24 and relative atomic mass 52. Its main ore is chromite. The ore is fused with caustic soda or sodium carbonate in air when it is converted to sodium chromate. On acidifying it changes into sodium dichromate. The ore on strong heating alone or in the presence of carbon decomposes to give chromium (III) oxide. Finally, it is reduced to chromium metal by aluminium. The metal resists aerial corrosion but reacts slowly with dilute acids to liberate hydrogen. Its important compound dichromate is used as a strong oxidising agent.

Chromium is used in strong alloy steels and stainless steel and for chromeplating of iron articles. Chromeplating is done for the protection as well as decoration purposes.

CHROMIUM (II) OXIDE : CrO; Chromous oxide. A black powdery substance prepared by reacting chromium amalgam with dilute nitric acid or by ordinary air. On heating in a current of hydrogen, it is changed to chromium.

CHROMIUM (III) OXIDE : Cr_2O_3; Chromic oxide. A green crystalline substance, insoluble in water. It is prepared by heating chromium (III) salts and in concentrated alkali solutions to give chromites. It is used as a pigment in paints and glasses.

CHROMIUM (IV) OXIDE : CrO_2; Chromium dioxide. A black, insoluble solid prepared by heating chromium (III) hydroxide in oxygen or by the action of oxygen on chromium (VI) oxide. It is the least stable among the oxides of chromium.

CHROMIUM (VI) OXIDE : CrO_3; Chromium trioxide. Chromic anhydride. A red crystalline solid prepared by the action of cold concentrated sulphuric acid to ice cooled saturated solution of potassium dichromate.

It is a powerful oxidising agent particularly for organic compounds. When dissolved in water it forms chromic acid which is used as an oxidising agent as well as a cleansing liquid for laboratory glass apparatus. On heating, it is converted to chromium(III) oxide.

CHROMIUM POTASSIUM SULPHATE : *See Chrome alum.*

CHROMIUM STEEL : Any of the stainless steels which contains 8 to 25% chromium. A typical chromium steel has 8% nickel 18% chromium and 0.15% carbon. They are highly resistant to corrosion and are used to manufacture cutlery, ball bearings and parts of equipment.

CHROMOPHORE : A group of atoms in a molecule which is responsible for the particular colour in a dye.

CHROMOUS COMPOUNDS : Compounds in which chromium exists in a lower oxidation state of +2.

CHROMOUS OXIDE : *See chromium (II) oxide.*

CHROMYL CHLORIDE : CrO_2Cl_2; Chromium oxychloride. A deep red coloured liquid, obtained as a vapour when solid potassium dichromate and sodium chioride are heated with concentrated sulphuric acid. It may also be obtained by the action of concentrated sulphuric acid on chromium (VI) oxide dissolved in concentrated hydrochloric acid. With alkalies, it reacts to form yellow chromates. It is a powerful oxidising agent particularly for organic compounds.

CHRYSOLITE : *See serpentine.*

CINNABAR : HgS; A red mineral form of mercury (II) sulphide, the principal ore of mercury.

CINNMALDEHYDE : C_6H_5-CH=CH-CHO; 3-phenylpropenal.

CINNAMIC ACID : C_6H_5-CH=CH-COOH; 3–phenyl–propenoic acid. A white crystalline unsaturated aromatic carboxylic acid. It occurs in essential oils in the form of esters.

CIS-TRANS ISOMERISM : *See geometrical isomerism.*

CITRIC ACID : HOOC-CH_2-C(OH)-COOH-CH_2 COOH. A white crystalline tricarboxylic aliphatic acid which occurs in citrus fruits. It is used in the manufacture of various syrups, squashes and soft drink concentrates.

CITRIC ACID CYCLE : *See Kreb's cycle.*

CLAISEN CONDENSATION : A reaction in which two molecules of ester condense to make a ketoester. For example, two molecules of ethyl acetate condense in the presence of sodium ethoxide to form ethyl acetoacetate. The reaction proceeds through the formation of carbanion, $CH_2COOC_2H_5$, produced by the action of sodium ethoxide on ethyl acetate and so the mechanism is similar to that of aldol condensation. The carbanion thus produced may attack another molecule of ester to make an intermediate anion which finally decomposes to give the products.

CLARK CELL : A voltaic cell containing an anode made of zinc amalgam and a cathode of mercury both dipped in a saturated solution of zinc sulphate. It was initially used as a standard cell which gave an emf of 1.4345 volts. Weston standard cell has now replaced it. Clark cell was named after British scientist H. Clark.

CLARK PROCESS : *See hardness of water.*

CLATHERATE : Cage compound or enclosure compound. A substance in which small molecules (guests) are trapped within the site of a crystalline (host) substance. They are not true compounds as no real chemical bonds are formed. The forces of attractions between the two types of molecules are only weak Van Der Waal's forces. Compounds are formed when the host compound is crystallised in the presence of guest molecules. Clatherates break to release the guest molecules either by heating or by dissolving in a solvent. Quinol and ice both form such compounds with sulphur dioxide and xenon. The size of guest molecules should be appropriate so that they just fit in the site formed by the host molecules.

CLAUDE PROCESS : A method for liquefying air on a commercial scale. Compressed air when expanded adiabatically, cools. This cool air is fed to a counter current heat exchanger where it reduces the temperature of the next lot of compressed air. The same air is recompressed and used again and after a number of repeated cycles, it liquefies. This process was developed by the French scientist George Cloude.

CLAUDETITE : As_4O_6; A mineral of arsenic.

CLAY : Naturally occuring aluminosilicates which form paste with water.

CLEAVAGE : The splitting of a crystal. It is also used for the breaking of a bond to divide the molecules into two parts. Cleavage may be homolytic or heterolytic.

CLEMMENSON'S REDUCTION : Reduction of carbonyl by zinc amalgam and concentrated hydrochloric acid to form corresponding alkanes, i.e. the conversion of $>C=O$ group to CH_2 (methylene group).

CLOSED CHAIN : A ring of atoms, e.g., cyclohexane or benzene.

CLOSE PACKING : The packing of spherical rigid particles (atoms, molecules or ions) so as to occupy the minimum space. In a single plane, each sphere is surrounded by six close neighbours in a hexagonal arrangement, while in space, a sphere is in close contact with twelve other spheres. It may be of two types. In hexagonal close packing, the spheres of the alternate layers are directly over one another and so the arrangement is ABABA. In cubic close packing the spheres of fourth layer are one above one another and so the arrangement is ABCABCA.

COAGULATION : The process of conversion of a colloidal solution into a precipitate. It is usually achieved by the addition of the required amount of an electrolyte to neutralise the charge stabilising the colloidal solution, e.g., ferric hydroxide sol is coagulated by sodium sulphate solution. Ions having higher charge and opposite sign to the sign of charge of sols are effective. Formation of river deltas is another example of coagulation of colloidal silt by the sea water. Coagulation may also be achieved by boiling, intermixing of oppositely charged colloidal solutions, persistent dialysis and continuous electrophoresis.

COAL : A brown black carbonaceous matter present below the surface of earth at various depths formed by the burial of big trees and other vegetation due to certain geographical changes. Wood originally contains about 40% carbon, so depending on carbonization, different varieties of coal are formed. These differ in carbon content, volatile matter and moisture. Out of peat, lignite, bituminius and anthracite, peat has the lowest carbon content (60%) while anthracite has the highest carbon content (90%). Bituminous is the most common variety of coal with 80% carbon which is used as fuel in household and in industry.

COAL GAS : A fuel gas produced by the destructive distillation of coal. It contains hydrogen 50%, methane 35% and carbon monoxide 8%. Along with the formation of coal gas, ammonia coal tar and coke are formed. It was earlier used as a fuel gas for a pretty long period. With the increased availability of natural gas, its use has now declined.

COAL TAR : A thick black liquid product of destructive distillation of coal. It is obtained as a byproduct of manufacture of coke. Tar is a source of a large number of aromatic compounds like benzene, naphthaline and their derivatives. It may be separated into various important compounds by fractional distillation. It is used for tarring of roads also. For long time, coal tar has been the source of chemical compounds for making drugs, medicines, paints, varnishes, plastics, synthetic fibres, explosives and pesticides. These days most of these compounds are being prepared from petroleum and petroleum products.

COBALT : Co; A light grey transition metal belonging to 3d series (4th period) of the periodic table. It belongs to the iron family and it is a member of group VIII with an atomic number of 27 and relative atomic mass 58.933. It is ferromagnetic in nature. Its ores are cobaltite, smaltite and erythrite. Cobalt ores are roasted to the oxide and then reduced with coke or water gas. It is used to make various useful alloys. Alnico is a well known magnetic alloy. Several stainless steels and high strength alloys may be manufactured which are resistant to corrosion and oxidation at high temperature and so are used to make cutting tools. Cobalt-60, the radio isotope is an important cancer treatment agent. Cobalt salts are used to give bright blue colour to glass, tiles and pottery. It was discovered by G. Brandt.

COBALT (II) OXIDE : CoO; A pink solid prepared by the action of caustic potash solution to cobalt (II) salt solution. We actually get a precipitate of cobalt (II) hydroxide on boiling. This on heating in absence of air changes to cobalt (II) oxide. The compound is oxidised in air to Co_3O_4 and is reduced by hydrogen to the metal.

COBALT (III) OXIDE : Co_2O_3; A black solid produced by strong heating of cobalt nitrate. On heating it readily changes into Co_3O_4 which is a mixture of cobalt (II) and cobalt (III) oxides.

COBALT STEEL : A group of alloy steels containing 5% to 12% cobalt, 14 to 20% tungsten, about 4% chromium and 1 to

2% vanadium. They are hard, brittle and used in making high speed tools.

COCALINE : $C_7H_{21}O_4$; A colourless crystalline, stimulant present in cocoa.

CODIENE : $C_{18}H_2NO_3$; A colourless crystalline compound used in the treatment of coughs.

COENZYME : A small organic nonprotein molecule which enables enzymes to carry out its biocatalytic activity. Coenzymes usually participate in the substrate–enzyme interaction and so they are not true catalysts.

COENZYME A : A complex organic compound which acts in conjunction with the enzymes involved in various biochemical activities like oxidation of pyruvate. It consists of principally vitamin B (panthenic acid) adenine and a ribose - phosphate group.

COENZYME - Q : Ubiquinone; A group of related quinone-derived compounds which serve as an electron carrier in cellular respiration. They have side chains of different lengths in different types of organisms but their function is the same.

COFACTOR : They may be organic molecules or inorganic ions which activate the enzyme by altering the shape. They may actually participate in the chemical reaction.

COHERENT UNITS : A system of units (S.I) in which the derived units are obtained by multiplying or dividing together base units without involving numerical factors.

COINAGE METALS : A group of malleable metals of group IB of the periodic table. They are copper, silver and gold. The general configuration of these metals is $(n\text{-}1)d^{10}ns^{1}$. They have much higher ionisation potentials than the alkali metals. They are classified with transition metals and so exhibit variable oxidation states. They also form a large number of coordination compounds. They were used in making coins in the past either individually or in the form of their alloys.

COKE : A form of carbon produced during the destructive distillation of coal. It is used as a fuel and as a reducing agent in metallurgical operations.

COLCHICINE : An alkaloid used in genetics, cytology and plant breeding research and also in cancer therapy to check cell division.

COLEMANITE : $CaB_3O_4(OH)_3.H_2O$. An important ore of boron.

COLLAGEN : A colourless fibrous protein containing glycine and proline predominantly. It is found extensively in the connective tissue of skin, tendon, bones and cartilage.

COLLIGATIVE PROPERTIES : The properties of a solution which depend on the concentration of particles or the number of particles (molecules or ions) present in the solution. They do not depend on the nature of the solute. Colligative properties include relative lowering of vapour pressure, osmotic pressure, elevation of boiling point and depression of freezing point. With the help of the measurement of colligative properties, molecular masses of solution can be calculated.

COLLIMITOR : An arrangement of lenses and slits used to produce a parallel beam of radiations for spectrometers or other instruments.

COLLISION : When two particles, atoms or molecules come close and collide with each other, they are said to have undergone a collision.

COLLISION FREQUENCY : It is the total number of collision per unit volume per unit time. Collision frequency depends upon (a) how closely the molecules are crowded together, (b) how large they are,and (c) how fast they are moving, which, is in turn depends upon their weight and the temperature.

COLLISION THEORY : The rate of reaction depends on molecular collisions. All collisions need to be effective to form products, which are, associated with definite minimum energy called threshold energy. Rate of a reaction may be given as
Rate = collision frequency × energy factor × probability factor (orientation factor).

COLLODION : A nitrocellulose used as a base for lacquers in which a ketone or ester may be used as a solvent.

COLLOIDS : Thomas Grahm classified the substances as colloids (starch, gelatin) and crystalloids (sugar, sodium chloride). He defined colloids as the substances which when mixed with the solvent did not pass through animal membrane. Later on, it was realised that colloids were distinguished from true solution by the presence of particles which were larger in size and so could not pass through animal membranes. Colloids are heterogenous mixtures of two or more phases one of which is dispersion medium while the other is the dispersed phase.

Based on the physical states, the colloids are of different types. Sols are the colloids of solids in liquids. Emulsions are the colloids in which both the constituents are in the liquid phase. Gels are the colloids in which the dispersed phase is liquid and the dispersion medium is solid.

Colloids are identified by particular properties like Tyndall effect, Brownian motion, eletrophoresis, coagulation, etc. Examples are starch, gelatin in water (lyophilic sols); ferric hydroxide, arseneous sulphide in water (lyophobic sols); milk (emulsion); mist, smoke (aerosal).

COLLOIDAL MILL : A device for converting a suspension into its colloidal state. It involves the disintegration of larger particles of suspension to smaller particles of colloids. The method is used in pharmaceuticals, paint industry, etc.

COLORIMETRIC ANALYSIS : Quantitave analysis in which the concentration of solute is measured by the intensity of the colour by comparing it with the colours of standard solutions.

COLUMBITE : (Fe, Mn) $(Nb\ Ta)_2O_6$; The main ore of niobium.

COLUMBIUM : An earlier name of the element niobium.

COLUMN CHROMATORGRAPHY : A type of chromatography which employs a solid stationary phase (adsorbent) like alumina or silica gel packed in a vertical tube. The sample solution is poured at the top and the adsorbed fractions can be washed of the column (elution) and collected by using an appropriate solvent (eluent). A proper choice of the eluent will allow the component to be selectively removed from the column.

COMBUSTION : A chemical reaction of complete oxidation of a substance in excess air or oxygen, accompanied by the release of heat and light. These usually involve a complex free radical chain reaction. Examples include burning of hydrogen to make water, burning of methane to make carbon dioxide and water.

Common-Ion Effect : Dissociation of a weak electrolyte is suppressed by the addition of a strong electrolyte having one ion common with that of the weak electrolyte. For example, dissociation of ammonium hydroxide is suppresed by the addition of ammonium chloride due to common ammonium ions, resulting in the overall decrease in the concentration of hydroxyl ions. This effect is used in the qualitative analysis of mixtures.

Common Salt : Sodium chloride.

Complex : A compound in which ligands (molecules or ions) form coordinate bonds with the central metal atom or ion. A complex may be positively charged, negatively charged or neutral. Complexes formed are generally coloured and paramagnetic due to the presence of one or more unpaired electrons in the (n-1)d orbitals. Examples are $Ni(CO)_4$; $[Co(NH_3)_3Cl_3]$; $[Fe(CN)_6]^3$;$[Cu(NH_3)_4]^{++}$;$[Cu(H_2O)_6]^{++}$.

Complexometric Titration : The titration in which the reaction involves the formation of an inorganic complex. For example, EDTA titrations to estimate Mg^{++} or Ca^{++} ions.

Component : A separate chemical substance in a mixture in which no chemical reactions are taking place, e.g., mixture of ice and water has one component. The number of components in a chemical system is defined as the smallest number of substances which are required to describe the composition of each phase present in the system either directly or with the help of an equation, e.g., decomposition of calcium carbonate is a two component system.

Compound : A substance obtained by the chemical combination of atoms of different elements in which the ratio of the combining atoms remains fixed and specific. The constituents of a compound cannot be isolated by easy physical means. Examples are ammonia, caustic soda, urea etc.

Concentration : The amount of substance dissolved per unit volume of solution or in a specific weight of solvent. It can be expressed in several ways like normality, molarity, molality, mole fraction etc.

CONCENTRATION CELL : A cell which functions due to the difference in concentration of solutions of the same electrolyte surrounding two electrodes.

CONCRETE : The construction material made by the hardening of cement, sand and water.

CONDENSATION : The conversion of a substance from a gas to a liquid (or a solid) or from a liquid to a solid phase by cooling.

CONDENSATION POLYMERISATION : The reaction of combination of two or more molecules followed by the removal of some simple molecules like water, methanol or ammonia.

CONDENSER : A device used to convert vapours into its liquid phase.

CONDUCTING POLYMER : A type of organic polymer which conducts electricity like a metal.

CONDUCTIVITY : K; Also called specific conductance. It is the reciprocal of specific resistance or resistivity. It is defined as the conductance of 1 cm long conductor having an area of cross section of 1 cm^2 or the conductance of 1 cm^3 of the solution. Its units are ohm^{-1} cm^{-1}.

CONDY'S FLUID : A solution of calcium and potassium permanganates used as an antiseptic.

CONFIGURATION : The arrangement of electrons around the nucleus in an atom or the distribution of electrons in various orbits and orbitals. It may also be described as the arrangement of atoms or groups in a molecule.

CONFORMATION : A particular geometry of a molecule resulting from rotation of one part of the molecule over the other molecule. They are generally obtained by the free rotation of one carbon atom over the other, linked by a single bond. For example, ethane exists in eclipsed and staggered conformations while cyclohexane exists in chair and boat conformations

CONGENERS : Elements belonging to the same group in the periodic table.

CONJUGATE ACID (or base) : When a base takes up a proton it forms its conjugate acid and when an acid loses a proton it forms a conjugate base. This is based on the Bronsted-Lowry concept of acids and bases. A conjugate acid-base pair is related by a proton, e.g., HCl and Cl, CH_3COOH and CH_3 $COO,^-$ NH_4^+ and NH_3. If the acid is strong its conjugate base is weak and vice versa.

CONJUGATION : A system of alternate double and single bonds, e.g., 1,3-butadiene has conjugated double bonds. Conjugated dienes are generally more stable than those of isolated dienes due to the delocalisation of the π-electron cloud.

CONSERVATION OF ENERGY (LAW OF) : It states that in all processes occuring in isolated system, the energy of the system remains constant.

Energy may change its forms or it may be transferred from one to the other part of the system but overall sum of all forms of energy does not change. The law was proposed by Helmoltz. It is valid only when mass is also conserved.

CONSERVATION OF MASS (LAW OF) : It states that matter can neither be created nor be destroyed by physical and chemical change although physical state may change. It is, of course, difficult to ascertain in practice, as a part of mass may change into energy by Einstein mass energy relationship, $E=mc^2$.

CONSOLUTE TEMPERATURE : The temperature at which two partially miscible liquids become fully miscible. It is due to the fact that with the rise of temperature, miscibility of liquids increases.

CONTANTAN : An alloy containing 50-60% copper and 40-50% nickel. It has a high coeficient of resistance. It is used in resistance wire and thermocouples.

CONSTANT BOILING MIXTURE : *See azeotropes.*

CONSTANT COMPOSITION (LAW OF): The composition of a pure chemical compound is always definite and independent of the method of its prepartion. It may also be termed as the law of definite proportions. It was formulated by Proust. For example, in case of carbon dioxide, the ratio of carbon and oxygen by weight is always 3: 8 whatsoever may be its source.

CONTACT PROCESS : An industrial method for the manufacture of sulphuric acid. In this method, sulphur dioxide is oxidised by air to sulphur trioxide using platinum or vanadium pentoxide. The optimum conditions for the process are a temperature of 723K and a pressure of about 2 atmospheres. As the catalyst is highly sensitive, the gaseous mixture being passed over it should be free from dust and impurities particularly arsenic impurities. Sulphur trioxide is finally absorbed into sulphuric acid of desired strength by dilution.

Continuous Phase : Dispersion medium of colloids.

Continuous Process : An industrial proces in which raw materials are constantly fed into the plant. These react as they flow through the equipment to give a continuous flow of products. Such a process is relatively easy to automate and thus, the product formed is cheaper. The fractional distillation of crude oil is a common example.

Continuous Spectrum : A spectrum made of a continuous range of emitted or absorbed radiations. Hot solids produce such spectra in infrared and visible regions.

Cooling Tower : Towers used to cool water. It is used in coolers and condensers.

Coordinate Bond : A bond made between two atoms by the unequal sharing of electrons. Here one of the atoms is electron rich and acts as donor while the other one is electron deficient and accepts a lone pair of electron donated by the doner. Donor molecules are called Lewis bases and acceptors are called Lewis acids. The coordinate bond after its formation behaves as a covalent bond.

Coordination Number : It is the number of groups, molecules atoms or ions surrounding a given atom or ion in a complex, or the number of monodentate ligands which may form coordinate bonds with a central metal atom or ion in a complex, e.g., in an octahedral complex the coordination number is six.

Copolymerisation : The process of polymerisation involving two or more monomers. During the formation of copolymers, some simple molecules like water are elimentated, e.g., nylon-66 is a copolymer of adipic acid and hexamethylene diamine.

COPPER : Cu. A red brown transition metal of 3d series having atomic number 29 and relative atomic mass 63.55. It is malleable, ductile, excellent conductor of heat and electricity and coinage metal. It occurs in nature as copper pyrites ($CuFeS_2$), cuprite (Cu_2O), azurite ($2CuCO_3Cu(OH)_2$) and malacite ($CuCO_3$ $Cu(OH)_2$. The metal is very less reactive towards air, water and dilute acids but in moist air, it forms a characteristic green surface layer. Alloys of copper, like brass bronze, etc. are extensively used.

COPPERAS : Green Vitriol $FeSO_4 .7H_2O$.

COPPER (I) CHLORIDE : CuCl; Cuprous chloride. A white solid substances insoluble in water. It is prepared by boiling a solution containing copper (III) chloride, copper turnings and hydrochloric acid. It is formed as ($CuCl_2$) complex ion. On pouring it into water a white precipitate of copper (I) chloride is formed. It is covalent in character and structurally resembles diamond. In the vapour phase, it exists in climeric state. It is used in Sandmeyer's reactions for converting diazonium salt to chlorobenzene.

COPPER (II) CHLORIDE : $CuCl_2$; Cupric chloride; a yellow powdery susbstance prepared by dissolving copper(II) oxide or copper(II) carbonate in dilute hydrochloric acid. On crystallisation green coloured crystals of $CuCl_2.2H_2O$ are obtained. The anhydrous salt is obtained by passing chlorine over heated copper. Dilute solution of copper (II) chloride is blue due to the presence of $[Cu(H_2O)_6]^{++}$ ions.

COPPER GLANCE : A mineral of copper with the formula Cu_2S.

COPPER (II) HYDROXIDE : Prepared by adding aqueous caustic soda solution to an aqueous solution of a copper (II) soluble salt. It is a blue coloured solid.

COPPER (II) NITRATE : $Cu(NO_3)_2.3H_2O$; A blue deliquescent solid prepared by treating copper (II) oxide or carbonate with dilute nitric acid and crystallising the resultant solution. On heating it directly, copper (II) oxide, nitrogen dioxide and oxygen are formed. The anhydrous salt is formed when copper is treated with a solution of nitrogen dioxide in ethyl acetate.

COPPER (I) OXIDE : Cu_2O; Cuprous oxide. A red solid, insoluble in water, obtained by reducing copper (II) sulphate solution in an alkaline medium by some organic reducing agent like glucose aldehyde, etc. It undergoes disproportionation when dissolved in dilute sulphuric acid.

With concentrated hydrochloric acid, it forms $[CuCl_2]$ complex ion. It is used in glass industry to impart red colour to glass.

Copper (II) Oxide : CuO; A black solid, insoluble in water, obtained by heating copper (II) carbonate or nitrate. It is soluble in dilute acids to give copper (II) salts in solution form which are blue.

Copper Pyrites : $CuFeS_2$. *See chalcopyrites.*

Copper (II) Sulphate: $CuSO_4 .5H_2O$. A blue crystalline solid, soluble in water, prepared by dissolving copper (II) oxide or carbonate with dilute sulphuric acid. The solution crystallises to pentahydrate. On heating to 383 K it changes into $CuSO_4.H_2O$ and finally to anhydrous copper (II) sulphate at 423 K temperature. This white anhydrous salt decomposes above 473 K temperature. It is hygroscopic and turns blue on absorbing moisture. Aqueous solution of copper sulphate is acidic in nature due to cationic hydrolysis. It is also known as blue vitriol and is used in the preparation of Borduex mixture (a fungicide), in the electroplating, in textile dyeing and in the preservation of timber.

Coprecipitation : The contamination of precipitate by substances which are otherwise soluble in the solvent. For example, barium sulphate is precipitated by adding sulphuric acid to an aqueous solution of barium nitrate. The precipitate contains some barium nitrate also.

Cordite : A propellent for guns. It is prepared by mixing cellulose nitrate and nitroglycerine with some plasticisers.

Corn Rule : A rule of determining D, L configuration of optically active compounds with alpha aminogroup. In case of alpha amino acids, the molecule is imagined as being viewed along H-C bond between the hydrogen and asymmetric carbon atom if the clockwise order of other three groups is COOH, - R and $-NH_2$ the compound belongs to D- otherwise to L-series.

CORRIN : Vitimin B_{12}.

CORROSION : Reaction of a metal with air, moisture and other compounds leading to the formation of oxide, carbonate hydroxide etc., on its surface. Rusting of iron is a common example of corrosion. It is dependent on the tendency of the metal to change into its naturally occurring mineral state.

CORUNDUM : Al_2O_3; Emery; A naturally occuring form of aluminium oxide. In pure state it is colourless and transparent, but acquires variety of colours due to impurities. Ruby, (red) contains chromium, and sapphire (blue) contains iron and titanium as impurities. It is very hard (second to diamond) and chemically resistant to weathering. It is used in various polishes, abrasives and grinding equipments.

COULOMB-C: The S.I. unit of electrical charge. It is the amount of charge transferred by a current of one ampere in one second. The unit is named after the French scientist Charles de Coulomb.

COULOMETER : An instrument to determine the electrical charge using electrolysis $Q = M/Z$ where Q is the amount of charge, M is the matter released or deposited and Z the electrochemical equivalent.

COUMARIN : $C_9H_6O_2$; Colourless ketonic compound used in perfumery.

COUNTERS : An instrument for the detection and estimation of radioactivity.

COUPLING : A type of chemical reaction in which two molecules join together, e.g., formation of azo compounds by the coupling of an aromatic diazonium compound and a phenolic or aromatic amino compound.

COVALENT BOND : A chemical bond formed between two atoms by the equal mutual sharing of electrons. Conventionally, it is denoted by a small horizontal line between the atoms. Covalent bonding according to modern theory is based on the overlapping of atomic orbitals of two atoms—these orbitals must have nearly equal energies, proper orientation and electrons with opposite spins.

COVALENT CRYSTALS : A crystal in which the atoms are covalenty bonded to one another. Sometimes they are termed as macromolecular crystals. They are hard and have high melting and boiling points due to strong lattice forces. Examples are diamonds, corborundum, boron nitride.

Covalent Radius : It is defined as half the distance between two covalently bonded atoms in a homodiatomic molecule like hydrogen or chlorine. For other heteromolecules substitutional methods are used because the sum of covalent radii of two atoms is equal to the distance between two nuclei. Covalent radii can also be determined for multiple bonds. Ex. covalent radii for carbon are 0.077 nm for single bond, 0.0665 nm for double bond and 0.0605 nm for triple bond.

Cracking : Thermal decomposition of high boiling fractions of petroleum to low boiling fractions. Molecules having larger number of carbon atoms break and form molecules having lesser number of C-atoms. During cracking saturated hydrocarbons break to form lower alkanes, alkenes and cyclic hydrocarbons. Catalytic cracking is a similar process in the presence of a catalyst which works at a lower temperature. Cracking is used to manufacture synthetic petrol, oil gas and petrol gas.

Cream Of Tartar : *See potassium hydrogen tartarte.*

Creosote : A dark coloured liquid mixture of phenols and cresols obtained by the refining of coal tar. It is used to preserve timber.

Cresols : Methyl phenols; A mixture of three types of cresols (ortho, meta and para) may be obtained by the distillation of coal tar. They are used as germicides and antiseptic.

Cristobalite : A mineral from of silicon (IV) oxide.

Critical Point : The conditions of pressure and temperature under which a liquid in a closed vessel remains indistinguishable from its vapour phase. For instance critical point for carbon dioxide is at 304.1 K and 73 atmosphere pressure.

Critical Pressure : It is the pressure of a liquid at its critical state or the lowest pressure required to cause liquifaction of a gas at its criticial temperature.

Critical Temperature : The temperature above which the given gas can not be liquefied even at high pressures. Gases like carbon dioxide, ammonia, chlorine have critical temperatures much above room temperatures, while gases like oxygen, nitrogen, hydrogen etc. have very low critical temperatures.

Critical Volume : The volume of a definite mass (usually one mole) of substances at its critical temperature and critical pressure.

Cross Linkage : A short side chain or branch or an atom which links two longer chains in a polymer.

Crotonic Acid : $CH_3CH = CH\text{-}COOH$; But-2–enoic acid; exists in the form of colourless needles in two geometrical isomeric forms, cis and trans.

Crucible : A dish or a vessel used for heating a substance.

Crude Oil : Petroleum–a naturally occurring mixture of hydrocarbons found below the rocky strata of the earth.

Crum-Brown-Gibson Rule : According to this rule, a group already present on the ring directs the incoming group either to meta position or to ortho, para position. If the hydride of AHA is directly oxidised to HOA, group A will be meta directing and otherwise it is ortho para directing.

Cryohydrate : A eutectic mixture of ice and some other substance (ionic salt) obtained by freezing a solution.

Cryohydric Point (Eutectic point) : The minimum freezing point for a set of components.

Cryolite : Na_3AlF_6 ; A white or colourless crystalline mineral of aluminium. It is mainly used for the extraction of aluminium from bauxite ore or for the electrorefining of aluminium.

Cryoscopic Constant : Depression of freezing point, decrease or lowering in the freezing point of a liquid when 1 mole of a non volatile solute is dissolved. It is denoted by K_f and its units are K Kg mol^{-1}

Cryostat : A container used to maintain low temperature. Dewar flasks are a common example.

Crystal : A solid substance that has a regular arrangement of particles, atoms or ions. It has fixed angles between its faces and definite ratio of edges.

CRYSTAL FIELD THEORY : A theory to explain the structures and properties of inorganic complex compounds. According to this theory, the central metal atom or ion is considered as a point positive charge and the ligands are considered as a point negative charges. The bonds between metal and ligands are considered as electrostatic or ionic in nature. The theory usually explains the spectra magnetic properties and colours of complxes.

CRYSTAL LATTICE : The regular pattern of atoms, ions or molecules in a crystalline substance. It is produced by the repetition of a unit cell in three dimensional space.

CRYSTALLINE : Substances with a regular internal arrangement of atoms, ions or molecules constituting the crystal.

CRYSTALLITE : A small crystal which has the tendency to develop into larger one. It is used to describe specimens which contain accumulations of many minute crystals of unknown composition and crystal structure.

CRYSTALLISATION : The process of forming crystals by the slow cooling of a solution saturated with a solute.

CRYSTALLOGRAPHY : The study of the formation of crystals, its structure and properties.

CRYSTOLLOIDS : *See colloids.*

CRYSTAL SYSTEM : A classification of crystalline solids based on the shapes of their unit cells. They are of seven types depending on the edges and the angles between faces.

CUBIC : Denoting a crystal in which the unit cell is a cube.

CUBIC CLOSE PACKING : *See close packing.*

CUMENCHE : $C_6H_5(CH_3)_2$; Isopropyl benzene. It is prepared by passing benzene vapours and propene over heated phosphoric acid catalyst. It is used for the industrial preparation of phenol.

CUPELLATION : A method used for the separation of noble metals like silver and gold from impurities like lead. The impure metal is heated strongly in a dish called cupel with a blast of hot air. Lead is oxidised to oxide which is carried away by the blast of air.

CURPRAMMONIUNM ION : $[Cu(NH_3)_4]^{++}$; The deep blue coloured tetramine copper (II) ion formed by dissolving a copper (II) salt in ammonia. It is used in the manufacture of rayon.

CUPRIC COMPOUNDS : Compounds containing copper (II) ions, e.g, cupric nitrate, cupric chloride, etc.

CUPRITE : A red mineral of copper, Cu_2O, an important ore of copper.

CUPRONICKEL : A group of copper-nickel alloys containing 20-30% nickel. they are very malleable and corrosion resistant. They are used in condenser tubes. Those with 25% nickel are used in coinage.

CUPROUS COMPOUNDS : Compounds containing copper (I) ions, e.g, cuprous chloride, cuprous nitrate, etc.

CURIE : Ci; A unit of radioactivity. It is the number of disintegrations produced by one gram of radium. It may also be defined as the amount of a given radioactive substance that produces $3.7x10^{10}$ disintegrations per second.

CURIE POINT : The temperature at which the ferromagnetic compound loses its magnetic property and becomes paramagnetic. In case of iron, Curie temperature (point) is 1033K and for nickel, it is 629K.

CURIUM : Cm; A highly poisonous radioactive, silvery, transuranic metal. Its atomic number is 96 and atomic mass of the most stable isotope is 247. It was discovered by G.T. Seaborg. It may be obtained by bombarding Americium-241 with neutrons. Curium isotopes have been used in thermoelectric power generators.

CYNAMIDE : (a) An inorganic salt containing the ion (CN2)2, e.g, calcium cyanamide. (b) A colourless, crystalline solid NH_2CH prepared by the action of carbon dioxide on hot sodamide. On hydrolysis, it changes to urea.

CYANIDE : (a) A salt containing CN^{-1} ion. (b) Nitriles or organic cyanides which produce carboxylic acids on hydrolysis. (c) A metal coordination complex formed with cyanide ions.

CYANIDE PROCESS : A method of extracting gold from its ores. The method is based on the principle of the leaching of ore with very dilute solution of potassium cyanide. Gold dissolves to make potassium aurocyanide, $KAu(CN)_2$, The complex thus formed is reduced with zinc.

Cyanine Dyes : A class of dyes that contain - CN = group linked with two heterocyclic rings containing nitrogen. They acts as sensitisers in photography.

Cyanocobalamin : *See vitamin B complex.*

Cyanogen : $(CN)_2$ or C_2N_2 ; A poisonous, colourless, pungent smelling gas which is soluble in water and alcohol. It is prepared in the laboratory by heating mercury (II) cyanide. It is manufactured by gas phase oxidation of hydrogen cyanide by air in the presence of silver catalyst. It is a pseudohalogen and is used as an intermediate for preparing various fertilisers.

Cyanohydrin : Organic compounds having a hydroxyl and cyanide group on the same carbon atom. Such compounds are prepared by the action of hydrogen cyanide on aldehydes or ketones. Acetaldehyde with hydrogen cyanide gives acetaldehyde cyanohydrin while acetone gives acetone cyanohydrin. On hydrolysis, they form 2-hydroxy carboxylic acid, e.g., acetaldehyde cyanohydrin produce lactic acid (2-hydroxy propanoic acid)

Cyanuric Acid : A white crystalline solid which is a trimer of cyanic acid $(HNCO)_3$. Structurally it is a six membered heteroring made of imide (NH) and carbonyl (CO) groups on alternate positions.

Cyclomates : Salts of the acid C_6H_{11} $NHSO_3H(C_6H_{11})$ – group is Cyclohexyl group. Sodium and calcium salts were used as sweetening agents in soft drinks.

Cyclenes : Cyclic hydrocarbons having at least one double bond.

Cyclic Compounds : Describing a compound having a ring of atoms. In homocyclic compounds all the atoms in the ring are of same element, e.g., carbocyclic ring contains carbon atoms only as in benzene, cyclopentane and cyclohexane. In a heterocyclic compound one or more atoms are different and are called hetero cycles, e.g., pyridine (C_5H_5N).

Cyclic Process : A process in which a given system undergoes several changes but finally returns to its original state.

CYCLISATION : A reaction involving the conversion of an open chain compound to a cyclic compound.

CYCLO : Prefix denoting a cyclic compound, e.g., cyclopentane, cyclohexane, etc.

CYCLOALKANES : Cyclic saturated hydrocarbons. They have a general formula C_nH_{2n}. They are less reactive than ordinary alkanes.

CYCLODIENE : 1, 4 - Dione (benzoquinone; quinone) ; A yellow solid, $C_6H_4O_2$, used in making dyes and in setting quinhydrone electrode to measure pH of a solution.

CYCLOHEXANE : C_6H_{12}; A colourless, liquid cycloalkane which occurs in petroleum. It is made by the hydrogenation of benzene in the presence of finely divided nickel at 423 K. It is used as a solvent and paint remover. On oxidation, it changes into adipic acid. The cyclohexane molecule may adopt two conformations–boat and chair, of which the chair form is more stable.

CYCLORITE : A highly explosive nitro compound, (CH_2HNO_2). In its ring structure, $(-CH_2)$ groups and nitrogen atoms are at alternate positions. It is obtained by nitrating $C_6H_{12}N$, urotropine hexamethylene tetramine. Cyclorite is a powerful explosive used for military purposes.

CYCLOPENTADIENE : C_5H_6; A colourless liquid having two double bonds obtained as a by–product during the fractional distillation of coal tar. The compound is not aromatic. But cyclopentadienyl anion $C_5H_5 \bar{y}$ is aromatic as it has three pair of delocalised pi-electrons. It combines with iron to form ferrocene, $(C_5H_5)_2$ Fe.

CYCLOTRON : A device used to accelerate projectiles like protons, neutrons etc.

CYSTEINE : A compound present in body proteins and is oxidised to cystine when exposed to air.

CYSTINE : Present in abundance in the proteins of skeletal and connective tissues of animals and in hair and wool.

CYTIDINE : A nucleoside comprising one cytosine molecule linked to D–ribose sugar molecule. Cytidine phosphates like CMP, CDP and CTP participate in several biochemical reactions particularly in the synthesis of phospholipids.

Cytochromic : A group of proteins with an iron containing group which form part of the electron transfer chain. Electrons are transferred by reversible changes in the iron atoms between the iron (II) and iron (III) states.

Cytosine : A derivative of pyrimidine. It is an important component of DNA and RNA.

Dakin's Solution : A solution of sodium hypochlorite (NaOCl) with 0.5% available chlorine neutralised with a weak acid like boric acid. It is used as an antiseptic lotion.

Dalton's Atomic Theory : A theory which makes the basis of atomic structure and the laws of combination, proposed by British scientist John Dalton. It states that -

(a) Atom is the smallest indivisible particle of an element.

(b) Atoms of an element are all alike but differ from the atoms of other elements.

(c) Atoms can combine with one another in a simple whole number ratio to make compound atoms (molecules).

(d) Compound atoms of a substance are also alike and different from those of other compounds.

(e) Atoms can neither be created nor destroyed.

The theory was used to explain various laws of chemical combination such as law of definite proportions, law of multiple proportions, law of conservation of mass, etc.

Dalton's Law Of Partial Pressures : The pressure exerted by a mixture of non reacting gases in a container is equal to the sum of the partial pressures of the components. Partial pressure means the pressure exerted by each individual constituent gas when filled alone in the same container at the same temperature. The law is obeyed by ideal gases. It is used to calculate the pressure of a dry gas from the pressure of a moist gas collected over water.

Daniel cell : A primary electrochemical or galvanic cell set up by copper in contact with copper (II) sulphate as cathode and zinc amalgam in contact with zinc (II) sulphate as anode. Oxidation takes place at anode while reduction takes place at cathode, and so the electrons flow from zinc anode (negative) to copper cathode (positive). In order to prevent intermixing of two electrolytes, separate containers connected with a salt bridge may be used or one of the solutions (usually zinc (II) sulphate) should be taken in a porous pot. Emf of the cell is 1.10 V. If dilute sulphuric acid (IM) is used, emf is found to be 1.08V. The cell is represented as :

$Zn(s) \mid Zn^{++}(1M) \mid\mid Cu^{++}(1M)$ G $\mid Cu(s)$

and the cell reaction is -

$Zn(s) + Cu^{++}(aq) \rightarrow Zn^{++}(aq) + Cu(s)$

It is named after its inventor, a British scientist, John Daniel.

DATING TECHNIQUES : Methods to estimate the age of earth, rocks, archeological articles etc. It depends on the existence of some measurable change that occurs at a known rate like radio carbon dating, fission - track dating, potassium - argon dating, uranium - lead dating etc. There is another type, which depends on the existence of something that develops at a seasonally varying rate, as in dendrochronology.

DATIVE BOND : *See coordinate bond.*

DAUGHTER : (a) A nucleide formed by the radioactive decay of some other (parent) nucleide. (b) An ion or free radical produced by dissociation or reaction of some other (parent) ion or radical.

DAVY LAMP : A safety lamp developed by Humphrey Davy to detect the presence of methane in coal mines.

D-BLOCK ELEMENTS : It is the block of elements in the periodic table consisting of thirty three known elements. These are arranged in four rows such as 3d elements from scandium to zinc, 5d elements from yttrium to cadmium, 4d from lanthanum to mercury and 6d elements having actinium, kurchatovium and halmium. Last of the series (6d) has only three while the other three have ten elements each. The general configuration of these elements is $(n-1)d^{1-10}\ ns^{1-2}$. They are all metals and known as transition metals.

DDT (DICHLORO DIPHENYLTRICHLOROETHANE) : White crystalline compound prepared by the action of chloral and chlorobenzene in the presence of concentrated sulphuric acid. It is a powerful and effective insecticide. The compound is stable, accumulates in the soil, and so may reach the food chain. Restrictions have now been imposed on its use.

DEACON PROCESS : An earlier method to manufacture chlorine by the oxidation of hydrogen chloride in air at 723K using copper chloride as the catalyst.

DEACTIVATION : A process in which the activity of a substance is decreased, e.g., poisoning of catalysts like platinium.

DEAERATION : A process for the removal of gases from solvents by physical and chemical means.

DEAMINATION : A process involving the removal of an amino group from an organic compound. For example, deamination occurs in the liver by certain enzymes whereby the amino acid is decomposed to ammonia.

DE BROGLIE EQUATION : Based on the quality of matter, i.e., the particles, during their propagation, are associated with waves, de Broglie proposed a relation between the momentum of the particle and the wave length associated: -h/p, where is the wave length h is Planck's constant and p is the momentum (mv). Microscopic particles with very low values of momentum are associated with appreciably large waves, while bigger particles with higher momentum may be associated with negligibly small waves.

DEBYE : D; A unit of electrical dipole by two charges described as the dipole moment produced by two charges of 10^{-10} electrostatic unit and placed at 10^{-8} cm distance. This value is equal to 10^{-8}. In S.I. cum (or 1 Debye) In S.I. units, it equals 3.33564×10^{-30} coulomb metre.

DEBYE-HUCKEL THEORY : The increase of equivalent conductance of strong electrolytes with dilution can be explained by this theory based on asymmetry effect (or relaxation effect) and electrophoretic effect. A mathematical equation was given for the purpose as.

$$\lambda_C = \lambda_O - (A + B\lambda_O)\sqrt{2}$$

Deca-Da : A prefix in metric system denoting ten times, e.g. 1 decametre = 10 metres.

Decahydrate : A crystalline hydrated solid having ten molecules of water of crystallisation, e.g., Galauber's salt, $Na_2SO_4.10H_2O$.

Decanoic Acid : $CH_3(CH_2)_8COOH$; Capric acid: A white crystalline saturated carboxylic acid. Its esters are used in perfumes.

Decay : The spontaneous radioactive decomposition. or conversion of a radioactive nuclide into a a radioactive or stable nuclide. The decay is a first order reaction and the decay constant is given by $\lambda = \frac{2.303}{t} l_n \frac{N_o}{N}$ where N_o and N are the number of radioactive nuclei at the starting time and after time t respectively. Reciprocal of decay constant is called half life. Half life period of a radioactive substance is given by t (1/2) = 0.693.

Deci : d; A prefix used to denote one tenth or 10^{-1}, e.g., 1 dm, = 10^{-1} m.

Decomposition : A process in which a compound breaks into smaller components, compounds or elements.

Decreptition : A peculiar cracking noise produced on heating certain crystalline solids due to the loss of water of crystallisation.

De-Emulsification : A process by which an emulsion can be broken into two constituent immiscible liquids. The methods may be chemical, electrical or even mechanical.

Defect : An irregularity in the orderly arrangement of a constituent particle in a crystal lattice. A point defect consists either of missing atoms or ions causing a vacancy in the lattice (Schottkey defect) or the missing atom or ion has moved to a nearby interstitial site (Frenkel defect). Both of these defects are stoichiometric, but in some cases, the defects are non stoichiometric like metal excess or metal deficiency defect. If more than one point defects occur in a crystal, there may be a slip along a surface causing line defect.

In general, defects are caused by strain or radiation. Above zero Kelvin all crystalline solids have a number of defects. Conducting properties in semi-conductors are due to the defects.

Definite Proportions (Law Of) : *See law of constant proportions.*

Degassing : Removal of absorbed gases from liquids and solids.

Degenerate Orbitals : Describing various orbitals that have the same energy. For example, all the five d-orbitals of a particular quantum number are degenerate. In the formation of complexes, when ligands approach the metal atom of ion, degeneracy is lifted.

Degradation : A type of chemical reaction in which a compound decomposes into a simpler compound in steps, e.g., Hofmann's degradation of an amide to form an amine with one carbon atom less.

Degree : A division on a temperature scale, e.g., on celius scale 1 is the hundredth part of the difference between freezing point and boiling point.

Degree Of Freedom : (a) The minimum number of independent variables required to define the state of a system, e.g., a gas has two degrees of freedom. It is given by Gibb's phase rule, $F = c + p + 2$, where c is the number of components, p is the number of phases and F is the number of variables or degrees of freedom.

Dehydration : The reaction by which water is removed or the reaction in which a compound loses hydrogen and oxygen in the ratio 2:1, e.g., ethanol when passed over heated alumina gets dehydrated to ethene.

Dehydrogenase : An enzyme capable of removing hydrogen atoms in biological reactions. They are very important in driving the electron transport chain reactions of cell respiration.

Dehydrogenation : A process by which the hydrogen atoms of a molecule are removed. By this, unstauration in the molecule is increased, e.g., aromatisation of cyclohexane to benzene.

Deliquescence : Highly hygroscopic substance. A substance which may absorb such an extent of water from the atmosphere that the given solid dissolves in that.

Delocalisation : When the valence electrons of some molecules are not restricted to definite bonds between atoms but are spread over several atoms in the molecule, such electrons are said to delocalised. It occurs mainly when the molecule contains conjugated double or triple bonds. In case of 1,3-butadiene, π electrons are delocalised and spread over all the four carbons atoms. Similarly, in case of benzene, six π electrons are spread over the whole of the ring.

Delocalisation is the cause of aromaticity in benzene and in general it increases the stability of the molecule.

Delta Iron : One of the forms of pure iron. It is the body centered cubic form which exists above 1676 K.

Denature : (a) To add a toxic and unpleasant substance to ethyl alcohol to make it unfit for drinking. For this purpose methyl alcohol or pyridine and a little copper sulphate is added.

(b) To produce a structural change in a protein or nucleic acid to change its biological properties. It is effected by heat or some chemicals. When an egg is boiled, its protein is denatured.

Dendrite : A crystal that has branched in growth. Such crystals give tree like appearance.

Dendritic Growth : Growth of crystals in a branched manner.

Dendrochronology : An absolute dating method based on the growth rings of tree. Due to similar climates, the trees in the vicinity show same pattern of growth ring, and it is possible to apply this growth ring pattern to calculate the age of the specimen of wood. Fossil specimens accurately dated by this method have been used to make corrections to the carbon dating techniques. It is also helpful in the study of past climatic conditions and extent of pollution.

Denitrification : A process by which nitrates in the soil are reduced to molecular nitrogen which is finally released into air. The process is carried out by the bacteria *Pseudomonas denitrificas*.

DENSITY : The mass per unit volume of a given substance. In S.I. units, it is described in Kg m^{-3}.

DEPOLARISATION : The prevention of polarisation in a voltaic or galvanic cell. For instance, manganese dioxide (depolariser) is placed around the positive electrode of a leclanche cell to oxidise the released hydrogen.

DEPRESSION OF FREEZING POINT : The decrease in the freezing point of a liquid when a nonvolatile substance is added to it. The extent of decrease depends on the amount of substance added and not on the nature of substance, so it is a colligative property. It is given by $T_f = K_f m$, where T_f is the decrease in freezing point. K_f is the molal depression constant dependent on the nature of solvent and m is the molality. It is measured by the Beckmann thermometer. The property is used to determine the molecular masses of substances.

DERIVATIVE : A compound which is derived from some other compound, e.g., benzoic acid is the derivative of benzamide. They are used to confirm the identity and structure of a compound analysed by qualitative method.

DERIVED UNITS : A unit that is derived from fundamental S.I. units (by the combination of other units), e.g., coulomb is derived unit, as the quantity of charge transferred by one ampere in one second.

DESALINATION : The removal of salt from sea water to make water fit for drinking as well as for irrigation purposes. Desalination using solar energy has the maximum potential because shortage of fresh water is most acute in hot regions. The other methods may be evaporation, freezing, reverse osmosis, electrodialysis and ion exchange.

DESICCATOR : A container for drying the substances and to keep them free from moisture. They usually contain a drying agent like silica gel or anhydrous calcium chloride.

DESORPTION : The removal of absorbed gases from the surface of a solid substance.

DESTRUCTIVE DISTILLATION : Distillation in absence of air. It is a process of strong heating in the absence of air so that the complex organic substances break down into a mixture of volatile compounds which are condensed. Destructive distillation of coal gives coal gas, coal tar and coke.

DETERGENTS : A group of substances used for cleansing purposes. Detergents may be soaps and synthetic detergents. Their structures are similar. Both have a polar end which is hydrophilic, while the other end is non polar and is lipophilic. Polar ends remain dissolved in water while the non polar ends remain attached to oil or grease. By this, the miscibility of the two increases. Detergents are also said to be significant as they decrease the difference in the surface tension of two types of liquids or actually lower down the surface tension of water. Soaps are the oldest detergents. Their disadvantage lies in their capability to react with hard water to form white curdy precipitates. Soaps are sodium salts of higher carboxylic acids. With hard water they form calcium or magnesium salts of these acids which are insoluble in water. On the other hand, synthetic detergents like sodium dodecylbenzene sulphonate (anionic detergent having $CH_3(CH_2)_{11}$ $C_6H_4SO_2O$ ions) or catonic detergent like CH_3 $(CH_2)_{15}$ $N(CH_3)_3^+$ Br may work well even in hard water. Synthetic detergents are not biodegradable and so cause water pollution.

DETONATING GAS : A mixture of hydrogen and oxygen gases in the ratio 2:1 by volume explode to form water when ignited.

DEUTERATED COMPOUND : A compound in which one or more hydrogen (^1H) atoms are replaced by deuterium atoms (^2H).

DEUTERIUM : D or ^2_1H; The isotope of hydrogen that has atomic number one and mass number 2.0. Its nucleus contains one neutron also. It is present in water in the form of oxide, heavy water (D_2O). It is obtained by the electrolysis of heavy water. Chemically, it resembles hydrogen but its reactivity is lesser. Its physical properties are also slightly different from those of hydrogen.

DEUTERIUM OXIDE : *See heavy water.*

DEUTRON : Positively charged ion formed when deuterium loses an electron. D^+.

DETRIFICATION : Loss of the amorphous nature of glass due to crystallisation.

DEWAR FLASK : Thermos flask or vacuum flask. A double walled vessel of thin glass with the space between the walls evacuated and sealed. Such a container is used to store the substances at a particular temperature independent of the temperature of surroundings. The vacuum between the walls restricts loss of heat by conduction and convection. The inner surface of the glass is silvered (by normal silver mirror method) to prevent heat loss by radiation. It was devised by British scientist, Sir James Dewar.

DEWAR STRUCTURE : A representation of benzene in which the opposite carbon atoms of the hexagonal ring are joined together while the carbon atoms on the side may be linked by double bonds. There may be three dewar structures for benzene which are considered as canonical forms of the resonance hybrid of benzene.

DEWPOINT : The temperature at which a mixture of air and water vapour is saturated with respect to water vapour.

DEXTRAN : A polymer of glucose manufactured by the fermentation of sucrose. It is used as a protective agent for colloids.

DEXTRINS : Intermediates formed during the hydrolysis of starch to sugars. They are used as adhesives.

DEXTROROTATORY : Property of an optically active compound which rotates the plane of polarised light to the right side. It is denoted by a − or + sign.

DEXTROSE : Grape sugar, Naturally occuring glucose. It is dextrorotatory and belongs to D series of configuration. So the name dextrose is used for D(+) glucose. It plays an important role in the energy metabolism of living organisms.

D-FORM : Dextrorotatory form of an optically active compound.

DIAGONAL RELATIONSHIP : First element of any group in the periodic table may resemble diagonally in properties with the second element of next group. For example lithium resembles magnesium; beryllium resembles aluminium and boron resembles silicon. The cause of such resemblance in properties is due to the similarity in atomic radii and charge of size ratio of these elements. Some of the similar properties of lithium and magnesium are as follows, (a) both burn in air to form normal oxides, (b) both form nitrides with nitrogen at high temperature, (c) halides of both are covalent and soluble in ethanol, (d) carbonates of both break on heating to make oxides, (e) nitrates of both the metals decompose on heating to make oxides.

DIALYSIS : A method by which the particles of crystalloid (true solution) may be separated from those of colloids by selective diffusion through animal membrane, cellophone or ultrafilters. The method is used for the purification of colloidal solutions. Particles of true solution being smaller pass through the fine pores of such semipermeable membranes while the bigger particles of sols are checked. Cell membrane of living organisms are semi-permeable. Dialysis takes place naturally in the kidneys for the excretion of nitrogenous waste from blood. In case of kidney failure, the job of purification of blood may be done by artificial dialysers.

DIAMAGNETISM : Describing a property of substances which causes repulsion of the substance when placed in a magnetic field.

1,6 DIAMINOHEXANE (HEXAMETHYLENE - DIAMINE) $H_2N(CH_2)_6NH_2$; A colourless solid made from cyclohexane by its oxidation to hexanedioic acid, reacting with ammonia to form ammonium salt and heating in presence of phosphorous pentoxide to hexanedinitride. It is then reduced with hydrogen to give hexamethylene diamine.

The compound acts as one of the monomers for the manufacture of nylon-66.

DIAMOND : A crystalline allotrope of carbon. It is the hardest naturally occurring substance used for jewellery and drilling or cutting equipment. In diamond, the carbon atoms form a three dimensional network with each C-C bond of 154 pm length and at an angle of 109°28 with neighbouring bonds. All these carbons are tetrahedrally arranged. The bonds are strong and covalent in nature giving diamond its strength and hardness. Diamonds may be synthesised from graphite in the presence of catalysts under very high temperature.

DIASPORE : A mineral of aluminium.

DIASTASE : An enzyme which catalyses the hydrolysis of starch to maltose. It is present in the moist semigerminated barley seeds.

DIASTEROISOMERS : Stereoisomers that are not identical and not mirror images of one another. Diastereoisomers have different physical properties like boiling point, melting point, solubility etc., and so may be separated from one another by the usual physical means like fractional distillation and fractional crystallisation. 2,3–dichloropentane forms four stereomers which exist as four pairs of diastereoisomers.

DIATOMIC : A molecule which has two atoms, e.g., O_2, N_2, Cl_2, HBr etc.

DIAZINE : *See azine.*

DIAZO COMPOUNDS : Organic compounds which have two nitrogen compounds linked together. They include azo compounds, diazonium compounds and compounds like diazomethane (CH_2N_2).

DIAZOTISATION : A low temperature reaction of aromatic amino compounds with nitrous acid to form diazonium salts. The temperature required is between 273 and 278 K. Above this temperature, diazonium salts are unstable and decompose to phenols. Nitrous acid for the reactions is prepared in *situ* by the action of sodium nitrite and dilute hydrochloric acid. The reaction is of synthetic importance as a large variety of organic compounds may be synthesised through diazonium salts.

DIBASIC ACID : An acid that has two replaceable hydrogen atoms in its molecule. Such acids form two series of salts–an acidic salt and a neutral salt. For example sulphuric acid, oxalic acid, etc.

1,2 DIBROMOETHANE : $BrCH_2 - CH_2Br$; Ethylene bromide; A colourless liquid made by the addition of bromine molecule on ethene. It is used as an additive in petrol to remove lead when tetraethyl lead is being used as antiknocking agent.

DICARBIDE : *See carbide.*

DICARBOXYLIC ACID : A carboxylic acid having two carboxyl groups in the molecule, e.g., oxalic acid, adipic acid etc.

DICHLORINE OXIDE : Cl_2O; Chlorine monoxide. A strongly oxidising gas, orange in colour. It is prepared by the oxidation of chlorine by mercury (II) oxide.

DICHIROISM : The property of some crystals of absorbing light radiations selectively in one plane and allowing the vibrations to pass through this plane at right angles. Polaroid is a synthetic dichromic substance.

DICHROMATE (VI) : Cr_2O_7; A salt containg Cr_2O_7 ions. Acts as a a strong oxidising agent. Changes into CrO_4^- ions in alkaline medium.

DIELS-ALDER REACTION : A type of reaction between a conjugate diene and a dienophile (a compound having one double bond and one carbonyl group). Reaction is due to the electron relasing nature of diene and the electron withdrawing groups in dienophile.

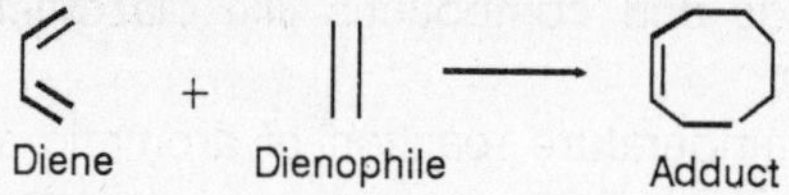

Diene + Dienophile → Adduct

The reaction is named after the German scientists Otto Diels and Kurt Alder.

DIENE : A compound having two carbon - carbon double bonds in its molecule. They may be conjugated 1,3 - butadiene or isolated like 1,4 - pentadiene or cumulated like 1,2 - propadiene.

DIENOPHILE : An unsaturated carbonyl compound capable of attaching itself with a diene. *See Diels-Alder reaction.*

DIETHYL ETHER : $C_2H_5OC_2H_5$; *See ethoxyethane*

DIETERIC'S EQUATION : $P(V-b) = RT\ e^{-a/RTV}$

Where a and b are Van der Waal's constant.

DIFFRACTION PATTERN : A pattern obtained on the photographic plate when a beam of X-rays or electrons is passed through a crystal. They are used to study the crystal structures.

DIFFUSION : Process of intermixing of substances due to random motion of constituent particles. Gases rapidly diffuse into one another as they are completely miscible into one another. Diffusion of gases obeys Grahm's law which relates the rates of diffusions of various gases with their molecular masses. Liquids and solutions also diffuse into one another but at a lesser speed.

DIFFUSION PUMPS : A mechanical device to reduce pressure with the help of oil or mercury vapours which carry out the gas molecules along with them. The gas molecules then condense on the colder walls of the pump.

DIGESTION : A process by which the size of the particles of a precipitate is increased so as to make it suitable for easy filtration. Usually it is done by heating the precipitate on a water bath particularly during gravimetric exercises.

DIHEDRAL : It is the angle made by taking a point on the line of intersection and drawing two lines from this point, one in each plane, perpendicular to the line of intersection.

DIHYDRATE : A crystalline compound having two molecules of water of crystallisation per molecule.

DIHYDRIC ALCOHOLS : Alcohols having two hydroxyl groups. *See diols.*

DIHYDROXYACETONE : $HOCH_2$ - $CO\text{-}CH_2OH$; A strong reducing agent, made by the oxidation of glycerol with Fenton's reagent.

1,2 - DIHYDROXYBENZENE : HOC_6H_4OH; Catechol; A colourless crystalline, phenol used as a developer (reducing agent) in photography.

2,3 - DIHYDROXYBUTANEDROIC ACID : *See tartaric acid.*

DIALTOMETER : An equipment used to measure small changes of volume of a solution or liquid.

DILUTION LAW : *See Ostwal's dilution law.*

DIMER : A compound or molecule made by the association of two identical molecules. Two molecules in the dimer may be held by hydrogen bonds as in carboxylic acid dimers or they may be joined by coordinate bonds as in aluminium chloride dimer, Al_2Cl_6.

DIMETHYLBENZENES : (Xylenes) - Compounds having two methyl groups substituted on the benzene ring, $(CH_3)_2\ C_6H_4$. Based on different positions they may be ortho, meta or para.

DIMETHYL GLYOXINE - $CH_3C = NOH$

$$\begin{array}{c} CH_3C = NOH \\ | \\ CH_3C = NOH \end{array}$$

Colourless crystalline solid soluble in ethanol. It is an important laboratory reagent to test nickel ions which form a red precipitate. It can be used for gravimetric estimations of Ni, Cu, Co, etc.

DIMORPHISM : *See polymorphism.*

DINITROGEN OXIDE : N_2O; Nitrous oxide. A colourless gas soluble in water and alcohol. It is commonly known as laughing gas. It is prepared by heating ammonium nitrate to about 523 K. It is relatively unreactive, inert to halogens, alkali metals etc. It is used as an anaesthetic and as an aerosol propellant.

DINITROGEN TETROXIDE : N_2O_4; A pale yellow liquid or a brown gas, dissolves in water to give nitric acid and nitrous acid. It may be formed by heating copper and concentrated nitric acid or by heating metal nitrates. N_2O_4 in the gas phase remains in equilibrium with NO_2. Liquid N_2O_4 is used as a non aqueous solvent (aprotic but ionising). It self ionises (like water) to NO^+ and NO_3^-. N_2O_4 along with other oxides of nitrogen are emitted from the exhaust of combustion engines and so cause pollution.

DIOL : An alcohol having two hydroxyl groups per molecule, e.g., ethanediol ($HO\text{-}CH_2\text{-}CH_2\text{-}OH$).

DIOXAN : $C_4H_8O_2$; A colourless poisonous liquid organic compound made from ethane - 1,2-diol and is used as a solvent.

DIOXONITRIC (III) ACID : *See nitrous acid.*

DIOXYGENYL COMPOUNDS : O_2^+ ion is called dioxygenyl cation and the compounds containing such a cation like dioxygenyl hexafluoroplatinate $O_2^+PtF_6^-$ are called dioxygenyl compounds.

DIPHOSPHANE : Diphosphine - P_2H_4; A yellow coloured liquid, inflammable in air, formed by the hydrolysis of calcium phosphide Ca_3P_2. Phosphine(PH_3) is always contaminated with diphosphine.

DIPOLE MOMENT : μ; A measure of the polarity of a molecule or an individual bond. Mathematically, it is the product of charge and the distance between two charged ends in case of a bond or diatomic molecule, e.g., HF. Resultant of all the bond dipoles is taken as the molecular dipole moment. With the values of dipole moment, geometries of the molecules may be confirmed. Molecules with zero dipole moment possess regular geometries, e.g., carbon dioxide is linear, carbon tetrachloride is tetrahedral, etc. Molecules having some definite values of dipole moment possess distorted geometries, e.g., water is angular and ammonia is pyramidal in geometry.

DIRECT DYES : A category of dyes derived from benzidine or its derivatives. They are generally azo dyes used to dye cotton, viscose rayon and cellulose fibres.

DISACCHARIDE : A carbohydrate made of two monosacoharide units, e.g., sucrose (made of a glucose and a fructose molecule), maltose (made of two glucose molecules).

DISILANE : Si_2H_6; A compound resembling ethane. *See silanes also.*

DISINFECTANTS : Substances used to destroy harmful micro-organisms.

DISLOCATION : A line defect. *See defect also.*

DISODIUM HYDROGEN PHOSPHATE (V) : $NaHPO_4$; A colourless crystalline solid soluble in water but insoluble in ethanol. It is used in treating water for boilers and in textile industry. It is also called disodium orthophosphate.

DISODIUM TETRA BORATE DECANYDRATE : *See borax.*

DISPERSE DYE : *See dye.*

DISPERSE PHASE : *See colloids.*

DISPLACEMENT REACTIONS : *See substitution reactions.*

DISPROPORTIONATION : A chemical reaction in which the same compound undergoes oxidation and reduction simultaneously. For example, formaldehyde disproportionates in the presence of concentrated caustic alkali to methanol and methanoic acid. Chlorine in aqueous caustic alkali disproportionates to a chloride (Cl) and a hypochlorite (ClO) ion.

DISSOCIATION : The breaking of a molecule into smaller molecules or ions. For example, hydrogen iodide may dissociate to hydrogen and iodine at high temperature.

$2\ HI(g) \rightarrow H_2(g) + I_2(g)$

The equilibrium constant in such a case is called dissociation constant. The term also applies to ionisation reactions of acids and bases in water.

$HA + H_2O \rightarrow H_3O^+ + A^-$

Equilibrium constant of such a reversible dissociation is called dissociation constant of acid (K_a) or acidity constant.

DISSOCIATION PRESSURE : When a solid dissociates to give some gaseous products, the pressure of the gas in equilibrium with the solid is called the dissociation pressure at a given temperature. For example, pressure of carbon dioxide when calcium carbonate is heated to a high temperature in a closed vessel gives its dissociation pressure.

DISTILLATION : The process of boiling a liquid and then collecting the condensate. Liquid thus collected is called distillate. The process is used for separation and purification of organic liquids. It may be of various types like, fractional distillation, steam distillation, vacuum distillation, etc.

DISTILLED WATER : Water free from dissolved salts and ions, with low conductivity. Repeated distillation in vacuum brings down the conductivity to 4.3×10^{-8} ohm^{-1} cm^{-1}, and this sample is called conductivity water. Distilled water is used in medicines, batteries and laboratories.

DISTRIBUTION LAW : At constant temperature, when different amounts of solute are allowed to distribute between two immiscible solvents in contact with each other, the ratio of the concentrations of the solute in two layers is constant at equilibrium (when the solute exists in same molecular form in both the solvents). The law is named after its inventor Nernst.

The law is used to determine the solubilities, molecular complexity and molecular mass or extent of association and dissociation. Further, it helps in finding the suitable solvent systems for the extraction. The law concludes that the higher the distribution ratio, larger is the amount extracted by a solvent, and by using the extracting solvent in larger number of lots of smaller volumes, the longer is the amount extracted as compared to that when the same volume of solvent is used in one lot.

DISTRIBUTION RATIO : K_D; It is the ratio of concentration of a solute in two immiscible solvents in contact with each other at a given temperature -$K_D = C_1/C_2$. It is also called distribution or partion coefficient.

DISULPHUR DICHLORIDE : S_2Cl_2; An orange coloured liquid prepared by passing chlorine over molten sulphur in the presence of iodine or some metal chlorides. In vapour phase, it exists as Cl-S-S-Cl. It is used as a solvent for sulphur and can form chlorosulphanes ($Cl-(S)_n-Cl$) which are of great value in vulcanisation.

DISULPHURIC ACID OR PYROSULPHURIC ACID : $H_2S_2O_7$; It is produced during the contact process when sulphur trioxide is absorbed in concentrated sulphuric acid. It is also called fuming sulphuric acid or oleum. It is used to prepare sulphuric acid of any desired strength and for the sulphonation of organic compounds.

DISYMMETRY : A property of the molecule responsible for its optical activity. It is also described as chirality of the molecules. Disymmetric molecules may exist in two enantiomeric forms, with the same capability of rotating the plane of plane polarised light but in opposite directions.

DITTRIONATE : $S_2O_6^{2-2}$; A salt containing $S_2O_6^{2-}$ ions or a salt of dithionic acid. It is generally obtained by the oxidation of sulphite by manganese (IV) oxide.

DIVALENT OR BIVALENT : Having a valency of two. For example, magnesium forms divalent cation, while sulphur forms divalent sulphide anion.

DL-FORM OR DL-MIXTURE : A mixture of dextro and laevo forms of a compound mixed in equal amounts so that the optical rotation of one is completely neutralised by the other. It is also called a racemic mixture.

D-LINES : Two close lines in the yellow region of the visible spectrum of sodium. Their wavelengths are 589 and 589.6 nm. They are used as a standard in spectrosocopy.

DNA : (Deoxyribonucleic acid) The genetic material of living orgnaisms, a major constituent of the chromosomes within the cell nucleus, playing a decisive role in determining hereditary characteristics. It is a nucleic acid made of two chains in which the carbohydrate is deoxyribose and bases are adenine, cytosine, guanine and thymine. Sugar is joined to the phosphate group. Two chains are spiralled over each other and linked through hydrogen bonds.

DNA is responsible for the control of protein synthesis in cells.

DOBERENIER'S TRIADS : A set of triads of chemically similar elements. When the triad was arranged in the order of increasing atomic mass, then the mass of the central element was approximately the average of the masses of other two. These were discovered by J.W. Dobereiner. The triads are now considered as the consecutive members of the groups of the periodic table, e.g., chlorine, bromine and iodine; lithium, sodium and potassium.

DODECANIOC ACID : $CH_3(CH_2)_{10}COOH$ (Lauric acid); A white crystalline fatty acid, present in the glyceride form in coconut and palm oil.

DODECANE : $CH_3(CH_2)_{10}CH_3$: A straight chain alkane obtained from petroleum.

DODECYLBENZENE : $CH_3(H_2)_{11}C_6H_5$; An alkyl benzene formed by the reaction between dodecene ($CH_3(CH_2)_9$ $CH=CH_2$) and benzene in phosphoric acid (Friedel Craft's reaction). Its sodium salt is used as a common detergent.

DOLOMITE : $CaCO_3$ $MgCO_3$; A carbonate mineral, usually white in colour.

DONOR : An atom, molecule or ion which is capable of forming a coordinate bond by donating a pair of electrons.

DOPA : (Dihydrophenylalanine) ; It is found in high levels in the adrenal glands and is the derivative of the amino acid tyrosine. It is used in the synthesis of dopamine, noradrenaline and adrenaline.

DOPAMINE : A catecholamine which is a precursor in the synthesis of noradrenaline and adrenaline. If serves as a neuro-transmitter in the brain.

D-ORBITAL : An orbital with $l=2$ and values of m between -2 and +2. D-orbitals may be five in number. They are double dumbell shape.

DOUBLE BOND : A bond made by the sharing of two pairs of electrons between two atoms. It is a combination of one sigma and one π bond. *See chemical bond for details.*

DOUBLE DECOMPOSITION : (Metathesis) ; A chemical reaction involving the exchange of ions, in aqueous solutions.

DOUBLE SALT : A salt made by the crystallisation of a mixture of aqueous solutions of two salts. Double salts retain their identity only in the solid crystalline state. On dissolving, they give rise to a mixture of cation and anions of both the salts. For example, alums are double salts of general formula, M_2SO_4 $M^2{}_2(SO_4)_3$. $24H_2O$ where M and M^1 are respectively monovalent and trivalent metal ions.

DOUBLET : A pair of associated lines in a spectrum, e.g., two lines making sodium D-lines.

DOWNS PROCESS : A process for extracting sodium by the electrolysis of sodium chloride mixed with some calcium chloride to lower the working temperature of molten electrolyte. It utilises graphite as, an anode and a cylinder of iron as a cathode. The two electrodes are separated by a hood so as to restrict the contact of sodium metal and chlorine liberated as the products of electrolysis.

DRY CELL : A cell in which the electrolytes are in the form of paste. Common dry cells are Leclanche cells in which ammonium chloride paste is used as electrolyte and carbon rod and zinc of the vessel act as electrodes.

DRY ICE : Solid carbon dioxide. It is used as a refrigerant. It is more convenient because at about 195K and standard pressure, it sublimes.

DRYING OIL : A natural oil which hardens on exposure to the air. These oils contain unsaturated fatty acids like oleic and linolenic acids which polymerise on oxidation. They are used in paints and varnishes. For example, linseed oil is a drying oil.

D-SERIES : *See absolute configuration.*

DULONG AND PETIT'S RULE : The product of the relative atomic mass and specific heat capacity of an element is a constant and nearly equal to 25 J mol^{-1} K^{-1}(or 6.4 cal). French scientists Pierre Dulong and A.T. Petit calculated the atomic masses of a large number of elements with the help of the rule. It also states that the molar heat capacity of solid elements is approximately equal to 3R where R is the universal gas constant.

DUMAS METHOD : (a) A method by which nitrogen may be estimated in a given organic compound. In this method, nitrogen of the compound is first converted to oxides of nitrogen with the help of copper oxide at a higher temperature. These oxides are then reduced by copper gauze to nitrogen gas which is collected in Schiffs nitrometer by the downward displacement of potash solution. Weight and percentage of nitrogen is calculated by measuring the volume of nitrogen evolved. (b) A method of determining relative molecular mass of a volatile liquid by finding the weight of vapours of known volume in a thin-glass bulb.

The methods are named after the French scientist Jean Baptiste Andre Dumas.

DUPLET : A pair of electrons in a covalent chemical bond.

DURALUMIN : A light weight alloy of aluminium containing 3-4% copper and traces of magnesium, managanese and silicon. It is used in the manufacture of the body of an aircraft.

DURIRON : An iron alloy containing 14% silicon, 2% manganese, 1% carbon and 0.1% sulphur. It is an acid resistant alloy and is used in making chemical plants.

DUTCH METAL : A zinc-copper alloy.

DYES : The coloured substances used to impart permanent colour to textiles, foodstuff, silk, wool, etc. The colour of a dye is linked to the nature of wave lengths absorbed and released in the visible region of light. If a dye absorbs radiations in the yellowish green colour, it will appear as violet. Previously dyes were mostly obtained from plants, e.g., indigo and alizarins, but now there are several synthetic dyes in use. Dyes are classified on the basis of their structures as, (a) Azo dyes; having $-N=N$ group, (b) Phthalein dyes; having phthalein group, e.g., phenolphthalein. (c) Indigoid dyes; oldest known dyes having indigoid group, e.g., indigo (d) Anthraquinone dyes; having an anthraquinone group, e.g., alizarin (e) Triphenyl methane dyes; having triphenylmethyl group, e.g., malachite green.

Dyes may also be classified on the basis of their application, like :

(a) Acid dyes; Sodium salts of sulphonic acids, e.g. orange-I.

(b) Basic dyes; having an amino group, e.g., aniline yellow and malachite green.

(c) Direct dyes; may be applied directly on the fabrics by dipping it in the aqueous solution of dye. These are water soluble, e.g., congo red.

(d) Vat dyes; one of the oldest dyes that imparts colour by oxidation and reduction. Indigo dye is a common example.

(e) Mordant dyes; effective only in the presence of some metal salts. Alizarin gives a rose red colour in the presence of Al^{3+} ions and blue colour with Ba^{++} ions.

(f) Disperse dyes; the dyes available in the form of fine

particles, as a suspension. When applied to the fibre they get dispersed on it, e.g., monoazodye and anthraquinne dye.

DYNE : Dyn; The former unit of force in c.g.s. system. It is equal to 10^{-5}N.

DYNAMIC EQUILIBRIUM : *See equilibrium.*

DYNAMITE : A type of explosive based on nitroglycerin. Originally invented by Alfred Nobel, it contained glyceryl trinitrate absorbed in Kieselguhr.

DYSPROSIUM : Dy ; A soft silvery metal belonging to the lanthanide series, having atomic number 66 and relative atomic mass 162.50. It occurs in apatite and gadolinite from which it is extracted by ion exchange method. Many of its isotopes occur in nature. Some artifically synthesised isotopes are also identified. Some of its alloys are used to absorb neutrons.

DYSTECTIC MIXTURE : A mixture of substances which has a constant maximum melting point.

E

EARTH : The planet that orbits the sun between the planets Venus and Mars. Earth is composed of different layers. A study of these layers and their constitutents has been made by Goldschmidt, Kuhn and Rittman. There is a metal core in the centre, called siderophile. It consists of elements like Mn, Fe, Co, Ni, Cu–mostly metallic and less reactive. Surrounding it is a rocky layer called sulphide layer or chalcophil. It is composed mainly of P, S, Zn, Ga, Ge, etc. The next zone or layer enveloping the chalcophil is the silicate layer or lithophil. It is made of chlorides, sulphates and carbonates of several metal like Li, Na, K, Rb, Cs, Mg, Ca, Sr, Ba etc. The layer which we actually see is atmophil or solid crust. It has halogens and noble gases as the main constituents. These layers have been formed by the slow cooling of magma coming out from deep inside the surface during volcanic eruptions that are taking place from time to time.

EARTH'S ATMOSPHERE : The gaseous mixture that surrounds the earth. It mainly consists of nitrogen (78.08%), oxygen (20.95%) carbon dioxide (.031%), neon (0.0018%), helium (.005%), krypton (.001%) and xenon (.00001%) when dry. In addition to water vapours, air may contain some other gases depending on a particular locality, e.g., sulphur dioxide, hydrogen sulphide, hydrocarbons, dust particles, smoke etc.

EBONITE : *See vulcanite.*

Ebullioscopic Constant : *See elevation of boiling point.*

Ebullition : The boiling or bubbling of a liquid.

Echelon : A device used in spectroscopy to study hyperfine line spectra.

Eclipsed Conformation : *See conformation.*

Edison Cell : *See nickel iron accumulator.*

Edta : $(HOOCCH_2)_2N(CH_2)_2N(CH_2COOH)_2$; ethylene diaminetetr acetic acid. A compound used as a chelating agent in coordination chemistry as it acts as a hexadenate ligand. It is also used to determine the strength of Mg^{++}, Ca^{++}, etc., by volumetric analysis.

Effective Atomic Number (EAN) : It is the total number of electrons available to the central atom from its own as well as electron pairs of ligands forming coordinate bonds. It is generally found that the effective atomic number is equal to the nearest noble gas. For example, in $Ni(O)_4$ complex, EAN of Ni comes to 36 which is the atomic number of Kr. This may not be be taken as a rule as it does not apply to a large number of coordination compounds.

Effervescence : The formation of gas bubbles in a liquid mixture due to a chemical reaction.

Efficiency : A measure of energy transfer. It is defined as the ratio of useful energy produced to the energy input in a system.

Efflorescence : The process in which a hydrated crystalline solid substance loses water forming a powdery amorphous compound at ordinary temperature in air.

Effusion : The passage of gas through a small hole or aperture. Effusion also obeys Graham's law of diffusion.

Eigen Function : An allowed wave function of a system in quantum mechanics. The associated energies are eigenvalues.

Einstein's Equation : The mass energy relationship proposed by Einstein in the mathematical form as $E = mc^2$, where E is the quantity of energy, m its mass and c is the velocity of light. The concept is applied to calculate the amount of nuclear energy based on mass defect or loss in mass. It can also explain the cause of stability of nuclei. It justifies the particle nature of light also.

Einsteinium : Es ; A radioactive transuranic element of the actinide series, having atomic number ·99. Mass number of its most stable isotope is 254. The element was first of all identified by A Ghiorso and associates in the debris from the first hydrogen bomb explosion.

Einstein's Law Of Photochemical Equivalence : Each reacting molecule must be excited by absorption of one quantum of light (or photon) in a photochemical reaction. It is true only for primary photochemical reaction. In secondary processes, a molecule once excited may lead to a chain reaction.

Elastomer : A natural or synthetic rubber which has the capability of being deformed under the influence of a force and regains its original shape when the force is removed.

Electrets : A permanently electrified substance which has oppositely charged ends. They resemble permanent magnets in many ways.

Electrical Double Layer : An electrical double layer developed at its interface between two phases of a colloidal system. Dispersed phase acquires one type of charge while the opposite charge is acquired by the dispersion medium.

Electric Arc Furnace : A furnace used in melting metals to make alloys in steel manufacture which uses the heat source as an electric arc. In the direct arc furnace like Heroult furnace, an arc is formed between the metal and an electrode while in the indirect arc furnance, the arc is made between two electrodes and then the heat is radiated on to the metla, e.g., Stassano furnace.

Electrochemical Cell : *See cell.*

Electrochemical Equivalent : Z; It is the amount of element liberated from a solution of its ions by the passage of 1 ampere current for 1 second (or by 1 coulomb charge).

Electrochemical Series : *See electromotive series.*

Electrochemistry : A branch of chemistry that deals with the study of conversion of chemical energy to electrical energy and vice-versa. Construction, working of electrochemical and electrolytic cells and the reactions involved are studied in it.

ELECTROCHROMATOGRAPHY : *See electrophoresis.*

ELECTRODE : A conductor that emits or absorbs electron in a cell. It may be described as a half cell also which shows either oxidation or reduction. Electrode at which oxidation takes place is called anode while the other at which reduction takes place is called cathode.

ELECTRODEPOSITION : The process of depositing one metal over the other by electrolysis, e.g., electroplating.

ELECTRODE POTENTIAL : E; A measure of the tendency of a half cell or electrode to lose electrons to form cations or gain electrons to form the metal. Tendency of an electrode to lose electrons is called oxidation potential and the tendency to gain is called reduction potential. Absolute value of electrode potential cannot be determined as a single half cell can never function without the presence of the other. Electrode potentials are determined by comparing with a normal hydrogen electrode of which value is arbitrarily fixed at zero volt. Standard reduction potentials are defined as the measurements of reduction potentials at 298 K temperature using 1M concentration of concerned ions. Numerical value of standard reduction potential is equal to that of standard oxidation potential but with the opposite sign.

ELECTRODIALYSIS : Dialysis under the influence of electric current, when the impurities are electrolytic in nature. This method is used to obtain pure water from saline water (desalination).

ELECTROLYSIS : The phenomenon of chemical changes taking place by the passage of electrical energy from an external source. In the process of electrolysis, cations move towards the cathode, gain electrons, get reduced and are liberated as neutral atoms or molecules. While anions by losing electrons get oxidised and are set free at the anode. This type of movement of ions towards oppositely charged electrodes is called electrolytic or ionilc conduction and is responsible for the flow of current through the electrolyte in the molten or dissolved state. Electrolysis is used for the extraction of reactive metals like copper, aluminium, silver etc., and electroplating, electro deposition etc.

ELECTROLYTE : Ionic or polar substances capable of producing oppositely charged ions in their solution or in the molten state. For example, sodium chloride, copper sulphate, dilute sulphuric acid etc. Liquid and molten metals conduct electrons due to the presence of free electrons and not ions. So they are not considered electrolytes.

ELECTROLYTIC CELL : A cell in which electrolysis occurs.

ELECTROLYTIC CORROSION : Corrosion that occurs through an electrochemical reaction, e.g., rusting of iron.

ELECTROLYTIC GAS : The highly explosive mixture of hydrogen and oxygen gases (2:1 by volume) obtained by the electrolysis of water.

ELECTROLYTIC REFINING : The method of purification of metals by electrolysis. It is applied to copper, aluminium, silver etc. For example, copper is refined by using copper (II) sulphate solution as electrolyte, impure copper metal as anode and a thin pure copper strips as cathode. The anode dissolves while copper ions from solution are reduced and deposited at cathode. Impurities are deposited at the bottom below anode.

ELECTROMAGNETIC SPECTRUM : The arrangement of different types of electromagnetic radiations in the increasing order of wave lengths or decreasing order of frequencies. The longest waves are radiowaves with wavelength of 1 m or more while smallest waves are gamma rays with wave length 10^{-10}m. Visible region of spectrum lies between 10^{-7} and 10^{-6}m.

ELECTROMETALLURGY : The use of electrical process in the separation of metals from their ores and the refining of metals.

ELECTROMOTIVE FORCE (E.M.F.) : It is the potential difference between the two electrodes, when no current is flowing in the circuit. It is the maximum voltage obtainable from the cell.

ELECTROMOTIVE SERIES : Electrochemical series. The arrangement of elements in the order of increasing standard reduction potentials. The values of standard reduction potentials are described in comparison to that of normal hydrogen electrode. The minimum reduction potential is of lithium or the half cell Li/Li^{++} (-3.05V), while maximum value is for flourine or F^-/F_2, Pt half cell (+2.87V). The electromotive series is used to determine the oxidising and reducing nature of the concerned elements, to determine the e.m.f. of the cells, in predicting the tendency of a metal to liberate hydrogen gas by reacting with dilute acids and in predicting the feasibility of a particular redox reaction.

ELECTRON : A fundamental subatomic particle with a mass of 9.10558×10^{-31} kg and a negative charge of 1.60219×10^{-19} C. Electrons are present in all atoms in the three dimensional region around the nucleus. They are distributed in various orbits and orbitals according to certain laws like Aufbau's Pauli's and Hund's laws.

ELECTRON AFFINITY : A; The amount of energy released when an electron enters the outermost orbit of an isolated gaseous atom. The number of electrons goes on increasing while moving from left to right along a period. Electron affinity to alkaline earth elements and elements of nitrogen family are nearly zero due to completely filled s-subshell or exactly half-filled configuration in the outermost orbit respectively. Electro-affinity of noble gases is zero due to the completely filled configuration.

ELECTRON CAPTURE : (a) The formation of negative ions by an atom or molecule by gaining electrons. (b) A radioactive change in which the nucleus gains an electron from the minor orbit of the atom (K-Capture).

ELECTRON DEFICIENT MOLECULES : A molecule in which there are fewer electrons forming the chemical bonds than required in usual bonds, e.g., borane. *See boranes for detail.*

ELECTRON DIFFRACTION : Diffraction of a beam of electrons by atoms or molecules. By this technique structures and shapes of molecules can be determined in the gaseous phase. A beam of electrons directed through a gas at low pressure produces a series of concentric rings on the photographic plate. The dimensions of these rings are related to the interatomic distances in the molecules. As the electrons undergo diffraction in a similar way to that of X-ray, electrons (particles) can also act as waves. It is employed to measure bond lengths and angles in gaseous molecules. Further, it is extensively used in the study of solid surfaces and absorption.

ELECTRONEGATIVITY : The tendency of an atom to attract the shared pair of electrons towards itself in a covalent bond. Fluorine is the most electronegative atom. There are two different ways of assigning values for electronegativity of an element. Electronegativity value is give by $E = (1 + A)/2$ where I is ionisation energy and A is electron affinity. Pauling scale of electronegativity is more common. These values are based on bond dissociation energies, using a scale in which flourine has a value of 4. Elements on the right side of periodic table have higher electronegativity values while those on left side have lower values. Difference in the electronegativity of two bonded atoms is the measure of polarity of a bond.

Electronegativity values of some of the important elements are given in the brackets.

Li(1.0), Be(1.5), B(2.0), C(2.5), N(3.0), O(3.5),
F(4.0), Na(0.9), Mg(1.2), Al(1.5), Si(1.8),
P(2.1), S(2.5), Cl(3.0), H(2.1)

ELECTRON MICROSCOPE : A type of microscope that uses a beam of electrons instead of a beam of light (as in the optical microscope). The transmission electron microscope has an electron beam sharply focused by lenses and a very thin visual image is formed. This image can be photographed. The scanning electron microscope can be used for thicker objects and forms a perspective image.

ELECTRON PROBE MICROANALYSIS (EPM) : A method of analysing a very small quantity of a compound (10^{-13}g). A very fine beam of electrons are focused on to the sample to produce the chracteristic X-ray spectrum of the elements present.

Electron Pair : Two electrons in an orbital with opposite spins.

Electron Spin : *See atom.*

Electron Spin Resonance (ESR) : A method similar to nuclear magentic resonance, but applied to unpaired electrons in a molecule (instead of the nuclei). It is a suitable method for the study of free radicals and transition metal complexes.

Electron Volt : eV; A unit of energy equal to the work done on an electron in moving it through a potential difference of one volt. 1 eV = 1.602×10^{-19}J.

Electrophile : An electron deficient ion or molecule which is capable of accepting electrons. They may either have a positive charge or an electron deficient atom. Common examples include NO_2^+; Cl^+; SO_3.

Electrophilic Addition : A reaction in which a small molecule adds to an alkene, an alkyne or some other electron–rich group. The mechanism of electrophilic addition is ionic through the formation of a carbonium ion or a carbocation. Addition of halogen acids on alkenes and alkynes is a common example. In case of higher alkenes, several isomeric products are possible. The particular isomer produced depends on the stability of the intermediate carbocation. It may also be based on Markonikoff's rule.

Electrophilic Substitution : A reaction involving the replacement of an atom or group of atoms in an organic compound with an electrophile as the attacking substituent. Aromatic hydrocarbons undergo such substitution reactions. Electrophiles for halogenation, nitration sulphonation and Friedel Craft's alkylation are X^-, NO_2^-, SO_3 and R^- respectively. The mechanism is through the formation of a resonance stabilised carbocation.

Electrophoresis : It is a technique used for analysing and separating colloids due to the movement of colloidal particles towards a particular electrode depending on the sign of charge on the colloidal particles. The rate of migration (movement) of the particles depends on the electric field applied, the amount of charge on the colloidal particles and the rate of migration (movement) of the particles. The technique is widely used in the study of mixtures of proteins, nucleic acids, carbohydrates, enzymes, etc. In pathological laboratories, it is used to determine the protein content of the body fluids.

ELECTROPLATING : The method of coating one metal over the surface of another by electrodeposition. The article to be plated serves as a cathode while the plating metal serves as an anode. A suitable electrolyte is used, e.g., for silver plating, dicyanoargentate (I) ions are used by taking $K(Ag(CN)_2)$ complex. It is used for decorative or ornamental purposes and for making the surface corrosion resistant.

ELECTROPOSITIVE : Describing elements that tend to lose electrons and form cations, e.g., Na, K, Mg, Ca etc.

ELECTROVALENT BOND : *See chemical bond.*

ELECTRUM : (a) An alloy of silver with gold containing 55-88% of gold.

(b) A German silver alloy containing 52% Cu, 26% Ni and 22% Zn.

ELEMENT : A substance having identical atoms, that is, the atoms having the same atomic number, although the mass numbers may vary. An element can not be decomposed into simpler substances. For example, hydrogen, oxygen, nitrogen, sulpher, iron, gold etc. There are 105 elements known so far of which 92 have been found to occur in nature, either in the native state or in the combined state. The rest of the elements have been synthesised in the laboratory. Elements may broadly be divided into metals (Na, K, Mg, Al, Cu, Mn, Pt etc.) non metals (H, O, N, Xe, He, S, P etc.) and metalloids (As, Sb, Bi etc.).

ELEVATION OF BOILING POINT : An increase in the boiling point of a liquid when a solute is dissolved in it. It is a colligative property and so does not depend on the nature of the solute but depends only on the number of moles dissolved in the given volume of solution. Mathematically, it is given by $\Delta T_B = K_b \times m$, where T_B is called the ebullioscopic constant and m is the molality of solution. K_B is constant and depends on the nature of solvent. It is measured by the Beckmann thermometer.

Elimination Reaction : A reaction in which a molecule decomposes and a smaller molecule is removed from it. For example, an alkyl halide on reacting with alcoholic potash solution changes into alkene eliminating halogen acid.

Elinvar : A nickel-chromium steel alloy with the composition 36% Ni, 12% Cr, and the rest iron. It also contains smaller amounts of W and Mn. Its elasticity does not change with temperature and so it is used to make hairsprings for watches.

Eluate : *See elution.*

Eluent : *See elution.*

Elution : The process of removing an adsorbed substance called adsorbate from an adsorbent by washing it in a liquid called eluent. The solution containing adsorbate dissolved in the eluent or solvent is called eluate. Elution may also be termed as washing the components through chromatographic column.

Emanation : The former name given to gaseous radon.

Emerald : A precious gemstone, green gem variety of beryl. They occur in nature in several countries and can also be synthesised.

Emery : Corundum mixed with spinel, magnetite or haematite and found as such in earth. It is used as an abrasive and polishing material and in the manufacure of certain concrete floors.

E.M.F. : *See electromotive force.*

Emission Spectrum : *See spectrum.*

Empirical : Representing a result based on experiements and observations and not on any theory.

Empirical Formula : *See formula.*

Emulsion : A liquid type of colloid, e.g., water in oil. Emulsions are always made in the presence of substance called emulsifier (a stabilising agent), e.g., a soap or a detergent.

Enantiomer : Enantiomorph; *See optical activity.*

Enantiotropy : The existence of different allotropes of an element at different temperatures. Sulphur exhibits enantiotropy. The phase diagram confirms that all these allotropes can coexist at a point. Above or below this temperature, only one will be more stable than the others.

ENCLOSURE COMPOUND : *See clatherate.*

ENDOTHERMIC : Representing a reaction in which heat is absorbed from the surroundings. If the reaction is allowed to take place in an insulated container, temperature of the system will fall. For example, dissolution of ammonium chloride in water, preparation of nitric oxide from nitrogen and oxygen.

END POINT : Equivalence point in volumetric analysis.

ENERGY : A thermodynamic function; A measure of the capacity of a system to do work. Its unit is joule, like that of work. It is of two types – potential energy which is the energy of a system as a result of its position, shape and state. It remains stored in the system itself. This includes gravitational, electrical, chemical and nuclear energy. Kinetic energy is the energy of motion. It is given $1/2\ mv^2$, where the mass of the body is m and it is moving with speed v. A vibrating or rotating molecule can have discrete vibrational and rotational energy. The internal energy or intrinsic energy of the body is the sum of kinetic and potential energy of all its atoms and molecules.

ENERGY LEVEL : A definite fixed energy that an atom, a molecule, an electron or the nucleus can have as per quantum theory. For example, electrons of an atom have different energy levels based on the orbitals to which they belong. An electron can be excited to a higher energy level also, but the increase of energy is either one quantum or its whole number multiples.

ENERGY PROFILE : A graphical diagram which exhibits the changes in the energy of a system during the course of a reaction. The graph is plotted between the potential energy of reacting particles against the progress of reaction. The energy profile for a reaction shows that the activation energy without the catalyst is much higher than that in the presence of a catalyst. The reacting molecules show an increase of energy till it acquires a maximum called threshold energy at which an activated transition state is formed. This intermedite may decompose to form the reactants or the products.

ENGEL'S SALT : *See potassium carbonate.*

ENOL : An organic compound having a carbon double bond as well a hydroxyl group on one of these carbons (-CH = (COH-)). *See keto-enol tautomerism.*

ENRICHMENT : The process of increasing the amount of a particular isotope in a mixture of isotopes. It is generally used to increase the percentage of a more useful radioactive isotope in the naturally occuring ore. For example, the proportion of U-235 is increased by this process in naturally occuring uranium.

ENTHALPY : H; A thermodynamic function. It is the sum of the internal energy and the product of pressure and volume of a given system, $H = U + _{P}{}^{V}$. Absolute value of enthalpy of a system in the given state can not be measured. In a chemical reaction carried out at constant pressure (atmospheric pressure) the change in enthalpy is given by

$$\Delta H = \Delta U + p\Delta V$$

The term $P\Delta V$ represents work of expansion or compression. ΔH is negative for an exothermic reaction and is positive for an endothermic reaction.

$$\Delta H = \Sigma H \text{ (Products)} - \Sigma H \text{ (reactants)}$$

Enthalpy change is a factor for predicting the spontaniety of a process.

ENTROPY : S; The quantity which measures the order and disorder of a system or the randomness of the system. More the disorder, more is the entropy of the system. Thus, entropy of vapour is more than that of the liquid, which in turn is more than that of the gaseous state. The symbol S for entropy was introduced by Clausius in the honour of Sadi Carnot. At absolute zero temperature there is complete molecular arrangement in the crystalline state. The entropy at this temperature is said to be zero. With the rise of the temperature, regularity of arrangment diminishes and so entropy increases.

Entropy is an extensive property. Change of entropy, ΔS i.e. $(S_2\text{-}S_1)$ of a system may be computed from the second law of thermodynamics. $\Delta S = \frac{Q}{T}$ when Q is the amount of heat change isothermally and reversibly at constant temperature T. It is expressed in JK^{-1} for a given system. If the amount of the substance is one mole, entropy should be expressed as JK^{-1} mol^{-1}. Entropy change is a factor used to predict the spontaneity of a reaction (like enthalpy change).

ENZYMES : Specially tailored protein molecules which catalyse the chemical reactions taking place in the living organism. Enzymes which act within the cell in which they are formed are called endoenzymes and those which act outside the cell are called exoenzymes. Several enzymes require some nonprotein cofactors in order to function. The molecule undergoing reaction (the substrate) binds to a specific active site on the enzyme to form a short lived intermediate. An enzyme increases the rate of reactions (sometime by a factor upto 10^{20}). The activity of an enzyme depends on the concentration of the substrate, pH of the medium and its temperature. Other molecules which compete for the active site cause inhibition. Most of the enzymes have their names similar to the corresponding substrates and usually they end in *-ase*. For example, maltase is the enzyme used to activate hydrolysis of maltose. Lactase is the enzyme that acts to break down lactose.

Several industrial processes use enzymes. For example, conversion of molasses to ethanol requires yeast which has several enzymes needed (invertase and zymase). Starch may also be converted to ethanol by fermentation which uses enzymes.

EPIMERISM : A type of optical isomerism in which a molecule has two chiral carbon atoms. Two optical isomers in this case, called epimers, differ in the arrangement about one of these chiral centers.

EPINEPHRINE : *See adrenaline.*

EPITAXY : Growth of a layer of one substance on a single crystal of another, such that the crystal structure in the layer is the same as that in the initial substance. It is used in making semiconductor devices.

EPOXIDES : Cyclic ethers; Compounds which contain an oxygen atom as a part of a three membered ring.

EPOXYETHANE : Ethylene oxide. A colourless inflammable gaseous compound C_2H_4O made by a catalytic oxidation of ethene. On hydrolysis, it produces ethane–1, 2–diol. On polymerisation, it gives epoxypolymers which are used to lower the viscosity of water as in fire fighting devices.

EPOXYRESINS : Synthetic resins produced by the copolymerisation of ethoxides with phenols. They can be hardened by addition of agents like polymines that form cross linkages. They are used in electrical equipment and in chemical industry due to their resistance to chemical attack. They are used as adhesives also.

EPSOMITE : Epsom salt; A mineral form of magnesium sulphate, $MgSO_4.7H_2O$.

EQUATION OF STATE : An equation that relates the pressue, volume, temperature and number of moles of the gas. Mathematically, for an ideal gas, it is $pV = nRT$. The real gases show large deviations from this ideal behaviour. A more accurate equation of state for real gases is given by Van der Waal, a Dutch scientist. It is given by

$$\left(\frac{P+ar^2}{v^2}\right)(V\text{-}nb) = nRT.$$

a and b are Van der Waal's constant. a denotes the intermolecular force of attraction while b represents the volume of the gas molecules. R is univeral gas constant which is 0.0821 L atmosphere IC-1 mol^{-1} or 8.314 JK^{-1} mol^{-1}.

EQUILIBRIUM : When a reversible reaction takes place in a close vessel, a state is attained after some time at which two opposite reactions–forward and backward proceed simultaneously and at the same rate. This state is called the equilibrium state. It is always dynamic in nature. All the observable properties remain unchanged. It may be attained in a physically reversible reaction like solubility, melting, boiling at constant temperature, as well as in a reversible reaction like formation of ammonia, hydrogen iodide or magnetic oxide of iron.

For example, in the reaction given below :

$3Fe(s) + 4H_2O(g) -- Fe_3O_4(s) + 4 H_2(g)$

Concentration of all the reactants and products acquire constancy although both the reactions are taking place at equal and opposite rates. The state of the equilibrium may only be changed when temperature, pressure or concentration of the components are changed.

It is explained in greater detail by Le Chatelier's principle.

EQUILIBRIUM CONSTANT : For a reversible reaction, $aA+bB - cC+dD$, it can be shown that the ratio of concentrations of the products to those of the reactants raised to the powers equal to the number of reacting molecules is a constant and is called equilibrium constant.

$$K_c = \frac{[C]^c [D]^d}{[A]^a [B]^b}$$

K_c represents equilibrium constant w.r.t. molar concentrations while square brackets represent the active masses of particular reactants and products. This expression is the mathematical form of the law of chemical equilibrium. For example, for the prepration of ammonia $N_2(g) + 3H_2(g) \rightarrow 2NH_3(g)$. Equilibrium constant is given by

$K_c = [NH_3]^2 / [N_2] [H_2]^3$

For the reactions involving gases, partial pressures are used in place of molar concentrations/active masses. For the reactions leading to the formation of ammonia the equilibrium constant w.r.t. partial pressure is given by

$$Kp = \frac{(pNH_3)^2}{(pN_2)\times(pH_2)^3}$$

It can be shown that for given reaction $K_p = K_c (RT) \Delta n$ where Δn is the difference in the number of moles of gaseous products and the number of moles of gaseous reactants. In this case, $\Delta n = 2 - (1+3) = -2$. The value of equilibrium constant depends on the temperature. If the forward step is exothermic, the equilibrium constant decreases with the rise of the temperature. On the other hand, for endothermic pressures, it increases. For accuracy, activities are used instead of molar concentrations.

EQUIPARTITION OF ENERGY : Proposed by L. Boltzmann and supported by James Clark Maxwell, it states that the energy of the gas molecules in a large sample under thermal equilibrium is equally divided for each degree of freedom being kT/2 where k is Boltzmann constant. It is not accurate but a good approximation.

EQUIVALENCE POINT : The point in a tiration at which the reactants have been mixed in equivalent proportions so that none is in excess. It is slightly different from end point, which is the observed point of complete reaction due to the effect of indicator as well as experimental errors.

EQUIVALENT PROPORTIONS (LAW OF) : Elements and compounds react together in the mass ratio of their chemical equivalent weights.

EQUIVALENT WEIGHT : It is the measure of the combining power of substances. The equivalent weight of an element is the number of parts by weight of an element which may combine with or displace from their compounds 1.008 parts by weight of hydrogen, 35.5 parts by weights of chlorine or 8 parts by weight of oxygen directly or indirectly. Mathematically it is given by Equivalent Weight = Atomic mass / Valency.

Equivalent weight of a substance depends on the manner in which it reacts. For some of the types of compounds, equivalent weight (E) may be described as –

E(salt) = Molecular mass / total positive valency of metal ions.

E(acid) = Molecular mass / basicity of acid.

E(base) = Molecular mass / acidity of base.

E(oxidant) = Molecular mass / no. of electrons gained by one molecule.

E (reductant) = Molecular mass / no. of electrons lost per molecule.

Erbium : Er; A soft silvery metal of the lanthanide series. It occurs alongwith other elements of the lanthanide series. It finds use in metallurgical, nuclear and glass industries. Its atomic number is 68 while its relative atomic mass is 167.26. It was discovered by C.G. Mosander.

Erg : A unit of energy in c.g.s. unit. 1 erg = 10^{-7} Joules.

Ergocalciferol : *See Vitamin D.*

Ergosterol : A sterol occurring in fungi, bacteria, algae and higher plants. When exposed to UV light, it changes into vitamin D_2

Erlenmeyer Flask : A conical flask having a narrow long neck.

Essential Amino Acid : An amino acid which must be present in the diet as it can not be synthesised by the organism. Essential amino acids in man are arginine, histidine, lysine, metheonine, isoleiucine, leucine, valine, phenylalanine and tryptophan. These are required for protein synthesis. These amino acids are essential for others animals as well.

Essential Element : An element required by living organisms for normal growth, development and maintenance. Elements found in organic compounds like C, H, N and O are essential, but some elements in the inorganic form like Ca, P, K, Na, Cl, S and Mg are also required. The trace elements occur at much less concentration and their requirement is also very low. Important among them are Fe, Mn, Zn, Mo, Cr and Si. Every element fulfils some specific metabolic role.

Essential Fatty : Fatty acids which must be present in the diet of animals. These include linoleic and linolenic acids. Deficiency of essential fatty acids may cause dermatosis weight loss, irregular oestrus etc.

Essential Oil : A natural oil with a distinct odour secreted by the glands of certain aromatic plants. Terpenes are the main constitutents and may be extracted from plants by steam distillation, followed by extraction with a solvent. They are used in perfumes and in medicines. Citrons oil, clove oil, rose and jasmine oils are common examples.

ESTER : An organic compound formed by the reaction between an alcohol and a carboxylic acid in the presence of concentrated sulphuric acid. Esters may also be formed by the acylation of alcohols and phenols by an acid anhydride or an acid chloride. Their general formula is RCooR'. Esters having simple and smaller alkyl groups are volatile and have a characteristic fruity smell.

ESTERIFIATION : The reaction of forming an ester from an alcohol and a carboxylic acid in the presence of concentrated sulphuric acid.

e.g. $CH_3\ COOH + C_2H_5OH \rightarrow CH_3COOC_2\ H_5 + H_2O$

The reaction is reversible or saponification (When carried out in presence of alkali).

ETHANAL : CH_3CHO; Acetaldehyde; A colourless, highly inflammable aliphatic liquid aldehyde which is made by the dehydrogenation of ethanol by its controlled oxidation, by the ozonolysis of 2-butene or by Rosenmund's reduction of ethanoyl chloride. It is used to prepare a large number of organic compounds.

The compound polymerises to a trimer called paraldehyde in the presence of dilute acid. The polymer is used as a sleep inducing drug. Addition of dilute acid below 273 K forms a tetramer called metaldehyde which is used as a solid fuel.

ETHANAMIDE : CH_3CONH_2; Acetamide; A colourless solid having a characterstic smell. It is made by the dehydration (by heating) of ammonium ethanoate or by the ammonolysis of ethanoyl chlorides, ethanoic anhydride or ethyl ethanoate.

ETHANEDIOIC ACID : *See oxalic acid.*

ETHANE-1,2-DIOL : $HOCH_2CH_2OH$; Ethylene glycol; A colourless viscous, hygroscopic liquid made by the hydrolysis of epoxyethane or by passing ethene through hydrogen peroxide solution or an aqueous alkaline KMn_2O_4 solution. It is a raw material for making terylene - a synthetic polymer. In western countries it is used as an antifreeze also.

Ethanoate : CH_3COO^-; Acetate; A salt or an ester of ethanoic acid.

Ethanoic Acid : CH_3COOH; Acetic acid; A colourless liquid with a characteristic sharp vinegar–like smell. It is manufactured by the oxidation of ethanol. Large quantities can also be prepared by the hydration of ethyne to ethanal which is finally oxidised by manganous acetate to get ethanoic acid. A dilute solution of the acid (5-8%) called vinegar is obtained by the oxidation of dilute ethanol solution by air in the presence of bacteria, *Mycoderma acetic*. The compound finds use as vinegar, as preservative and for the formation of ethanoic anhydride which is required for the manufacture of cellulose acetate.

Ethanol : CH_3CH_2OH; Ethyl alcohol; A colourless water soluble liquid with a peculiar smell. It is manufactured by the fermentation of sugar (from molasses) or starch using yeast. Ethanol obtained by this method has a strength of only 18% by volume as ethanol kills the yeast cells. Distillation of this dilute alcohol (wash) may produce an azeotrope containing 95.6% ethanol and 4.4% water (rectified spirit). Pure (absolute) alcohol may be obtained by removing the traces of water by means of drying agents like magnesium or calcium pieces. The main use of alcohol is in the manufacture of various beverages (intoxicating drinks). It is further used as an industrial solvent, as a fuel in the form of methylated spirit and as a starting material for several organic compounds.

Ethanoyl Chloride : CH_3COCl; Acetyl chloride; A colourless liquid with a pungent smell. It is prepared by the action of phosphorous (III) chloride or phosphorous (V) chloride or sulphur dichloride oxide on ethanoic acid. It is an acetylating agent, i.e., a compound to introduce CH_3CO–group. For example, by Friedel-Craft's reaction, it converts benzene to acetophenone.

Ethanoyl Group : CH_3CO–; Acetyl group.

Ethene : $CH_2=CH_2$; Ethylene; A colourless, unsaturated hydrocarbon gas, produced by cracking hydrocarbons from petroleum. It is now used as an important starting material for the manufacture of ethanol, ethandiol, ethanal, polythene, etc.

ETHENYL ETHANOATE : $CH_2=CH\text{-}OOC\text{–}CH_3$; Vinyl acetate; An unsaturated ester, made by the action of ethene and ethanoic acid. It is used for the manufacture of polyvinyl acetate.

ETHER : ROR ; An alkoxyalkane or ether is the compound having –O– linked to two alkyl or aryl groups. Usually they are volatile, inflammable liquids made by the dehydration of alcohols.

ETHOXYETHANE : $C_2H_5OC_2H_5$; Diethyl either; A colourless, inflammable, organic liquid, insoluble in water. It is made by the dehydration of ethanol or by Williamson's synthesis (action of ethyl chloride and sodium ethoxide). It is an anaesthetic and is also used as a solvent.

ETHYLACETATE : *See ethylethanoate.*

ETHYL ALCOHOL : *See ethanol.*

ETHYLAMINE : $C_2H_5NH_2$; aminoethane; ethanamine; a colourless volatile liquid made by the reaction of ethyl chloride and concentrated ammonia solution (in excess) at about 383 K. The compound formed is contaminated. with other amino compounds, and so, to get pure ethyl amines other methods like Hofmann's bromanide degradation or Gabriel's phthalimide synthesis may be used. It is used for the manufacture of several dyes.

ETHYLBENZENE : $C_6H_5\text{– }CH_2\text{–}CH_3$; A colourless inflammable liquid made from ethane and benzene by Friedel Craft's reaction. It is used for preparing styrene $C_6H_5\text{–}CH=CH_2$ (phenyl ethene) which is further used for synthesising several polymers.

ETHYL BROMIDE : *See bromoethane.*

ETHLENE : *See ethene.*

ETHYLENE OXIDE : *See epoxyethane.*

ETHYL ETHANOATE : $CH_3COOC_2H_5$; Ethyl acetate. A colourless inflammable liquid ester prepared by reacting ethanol and ethanoic acid or ethanoyl chloride or ethanoic anhydride. It is used as a solvent as well as in perfumery.

ETHYL GROUP : C_2H_5 group.

ETHYNE : HC=CH; Acetylene; A colourless, unsaturated hydrocarbon gas with a peculiar smell. It is prepared by the action of water on calcium carbide. It is manufactured by decomposing limestone to calcium oxide, which on heating strongly with coke produces calcium carbide. Ethyne is useful in the manufacture of ethanal, ethanol and ethanoic acid. As such, it is used for oxyacetylene welding and may be polymerised to give various important products. It is also used in the industry to manufacture chloroethene (a starting material for making PVC) and chloroprene.

EUCALYPTUS OIL : An oil extracted from the leaves of eucalyptus trees. It is used as an antiseptic and as an active agent in medicines for cold and flu.

EUDIOMETER : An apparatus for the determination of molecular formula of a hydrocarbon or other gaseous compounds. It consists of a graduated tube by which the changes in volumes of gases during burning and other chemical reactions may be measured. For the combustion of a hydrocarbon the following equation may be used to determine the number of C and H-atoms in its molecule.

$$C_x H_y(g) + \left(\frac{x+y}{4}\right) O_2(g) \rightarrow xCO_2(g) + \frac{y}{2} H_2O\ (l)$$

EUROPIUM : Eu; A soft silvery metal of the lanthanide series having atomic number 63 and relative atomic mass 151.96. It occurs in small amounts in bastanite and monazite ores. Eu-151 and Eu-153 are stable and occur in nature. Its main use is in a mixture of europium and yttrium oxides used as red phosphor in television screens. The elements were discovered by Sir William Crookes.

EUTECTIC : A mixture of two substances in such a proportion that it has the lowest freezing point. The minimum freezing point for a set of components is called eutectic temperature. Low melting point alloys are usually eutectic mixtures. If one of the components is water the eutectic is called cryohydrate.

EVAPORATION : The change of a liquid into its vapour phase at a temperature below the boiling point of the liquid. Evaporation occurs at the surface of a liquid and so causes cooling.

Exa : E; A prefix in the metric system to denote 10^{18}.

Excitation : A phenomenon of increase of energy of particle like nucleus, electron, atom, ion or molecule to a quantum state higher than that of its initial or ground state.The amount by which the energy is increased is called the excitation energy.

Exclusion Principle : (Pauli's) *See Pauli's exclusion principle.*

Exothermic : Describing a reaction in which heat is evolved or released. In such a process, energy of the products is lower than that of reactants. Sign of enthalpy change for these reactions is negative.

Explosives : Compounds which undergo a vigorous reaction evolving a large amount of heat and cause a sudden increase in pressure. Examples are gunpowder, cellulose nitrate, nitroglycerin, trinitrotoluene, trinitrobenzene, etc.

Extender : An inert substance added to a product to dilute it and to lower its cost. Some of the physical properites also change by the addition of extender. Substances to which additives are mixed are paints, rubbers, washing power etc.

External Indicator : An indicator used outside the reaction vessel, e.g., potassium ferricyanide in the titration of potassium dichromate against ferrous sulphate or Mohr's salt.

Extinction Coefficient : A characteristic property of a medium which absorbs light and is the reciprocal of the layer thickness measured in centimeters, in which the intensity of the incident radiation has been reduced to 1/10 of its original value.

Extraction : The process involving the removal of a soluble component from a mixture by a suitable solvent. The term is also used for the isolation of a metal from its ores.

F

Face central Cubic Crystal (FCC) : A crystal structure in which the unit cell has atoms, ions or molecules at every corner as well as at each face centre of the cube. It has a coordination number of 12. The structure is closely packed and made up of layers of atoms in which each particle is surrounded by six other particles arranged hexagonally. If the layers are designated A, B, C, the packing arrangement is represented by ABC ABC... Examples are copper and aluminium.

Fahrenheit Scale : A scale of temperature in which the freezing temperature of water is taken as 32° and the boiling point of water is taken as 212°. The scale is named after its inventor, a German scientist G.D. Fahrenheit. He considered zero as the lowest temperature obtained by mixing ice and common salt and took his own body temperature as 96°. The scale is still in use in medical sciences. It may be converted to the celsius scale by the equation :

$C = 5 (F - 32)/9$

Fajan's Rule : According to the rule, a chemical bond is never completely ionic or completely covalent when present between two different atoms. The magnitude of covalent character present in an ionic bond due to polarisation of the ion is characterised by these rules. A bond will be more covalent if, (a) the charge of the ions is high, (b) the positive ion is small or the negative ion is large, (c) the positive ion has an outer configration which does not resemble that of the nearest inert gas. It is named after the Polish–American scientist, K Fajan.

FARAD : F; The S.I. unit of a capacitance. It is defined as the capacitance when the charge is one coulomb and a potential difference of one volt is applied between its plates ($IF = 1CV^{-1}$). It is too large for most applications, the practical unit is microfarad ($IF = 10^6 \mu F$).

FARADAY OR FARADAY CONSTANT : It is given by $IF = N_A e$, where N_A is Avogadro's number and e is the amount of charge present on one electron. Its value is 9.648670×10^4 coulomb. It may also be defined as the amount of charge required for depositing 1 mole of monovalent metal on cathode during electrolysis.

FARADAY'S LAWS : The laws were initially proposed by Michael Faraday. The laws are state : (a) The amount of the substance deposited during electrolysis is directly proportional to the charge passed, i.e., $W = itz$, where w is the amount of substance deposited, i is the current passed in amperes, t is time in seconds and z is the electrochemical equivalent.

(b) The amount of chemical change produced in different substances by a fixed amount of electricity is proportional to the electrochemical equivalent of the substance.

Electrochemical and chemical equivalent of a substance are related by Faraday's constant.

$E = Z \times F$, where E is the equivalent weight, Z is the electrochemical equivalent and F is the value of Faraday's constant.

FATS : Triesters of glycerol and higher straight chains. In animals, they make the fatty tissues of the body. In plants, they occur chiefly in spores, seeds like cotton, sunflower, linseed and coats of fruits like palm and olive. They form food reserves for the plant embryo during germination and early growth.

FATTY ACIDS : *See carboxylic acids and essential fatty acids.*

F-BLOCK ELEMENTS : The f-block consists of elements of lanthanide and actinide series in the periodic table. They are characterised by the configuration $(n-2)f^{1-14}(n-1)d^{0-1}ns^2$. The total number of elements in this block is twenty eight. There are fourteen lanthanides (cerium-58 to lutenium 71) and fourteen actinides (thorium-90 to lawrencium-103). Actinides include transuranic elements (neptunium–93 onwards).

FEHLING'S TEST : A test to detect reducing sugars and aldehydes in solutions. It was introduced by the German chemist, H.C. Fehling. Fehling's solution consists of Fehling's A, aqueous copper (II) sulphate and Fehling's B, a mixture of sodium hydroxide and sodium tartarte. Equal volumes of these two solutions are mixed and heated with the given substance. A brick red precipitate of copper (I) oxide indicates the presence of aldehyde group or some other reducing group. Ketones give a negative test.

FELDSPARS : A group of silicate minerals. They have a structure in which the molecules are linked together with some metals ions occupying the large spaces in the framework. Their composition is expressed as a combination of four components.

a) anorthite $CaAl_2Si_2O_8$
b) albite $NaAlSi_3O_8$
c) orthoclase $KAlSi_3O_8$
d) celsian $BaAl_2Si_2O_8$

Feldspars form colourless, white or pink hard crystals.

FELDSPATHOIDS : A group of alkali aluminosilicate minerals that are similar in chemical composition to feldspars but are deficient in silica and rich in alkali metals.

FEMTO : f; A prefix used in the metric system to denote 10^{-15}. 1 fs = 10^{-15}s.

FERMENTATION : A process which resembles boiling. It is a type of anaerobic respiration occuring in some microorganisms, e.g., yeast. It is applied to convert sugar to ethanol and carbon dioxide.

FERMI : A unit of length used in nuclear physics. 1 fermi = 10^{-15} metre = 1 femto metre. It was named after the U.S. scientist Enrico Fermi.

FERMIUM : Fm; A radioactive metal of the actinide series having atomic number 100. Mass number of its most stable isotope is 257. The element does not occur in nature and can only be synthesised by artificial transmutations. It was first identified by A. Ghioroso in the debris from hydrogen bomb explosion.

FERRI COMPOUNDS : Compounds having iron in its +3 oxidatior state.

FERRIC CHLORIDE : *See iron (III) chloride.*

FERRIC OXIDE : *See iron (III) oxide.*

FERRIC SULPHATE : *See iron (III) sulphate.*

FERRICYANIDE : A compound containing the ions $[Fe(CN_6)^{3-}]$ hexacyanoferrate (III) anion.

FERRITE : Mixed oxide of the $MOFe_2O_3$ where M is a divalent metal such as cobalt, managanese, nickel, zinc etc. The ferrites are ceramic materials and act as non-conductors.

FERRIC ALLOYS : Alloys of iron made by smelting a mixture of iron ore and the metal ore (of some other metal). For example, ferrochromium, ferrovanadium, ferromanganese etc. They are used in manufacturing various types of steels.

FERROCENE : $Fe(C_5\text{-}H_5)_2$; An orange–red, crystalline solid prepared by adding cyclopentadienyl sodium (NaC_5H_5) to iron (III) chloride. In this compound two rings are parallel and Fe^{3+} ion is sandwiched between them. The bonding is between orbits of the rings and d-orbitals of the iron ion. Its formula is written as $Fe(C_5H_5)_2$, this means that all the five carbon atoms of the ring are bound to the metal. Ferrocene is an example of a π-bonded organometallic compound like dibenzene chromium.

FERROCYANIDE : A compound containing $[Fe(CN)_6]^{4-}$ complex ions called hexacyanoferrate (II) anion.

FERROELECTRIC : It is observed in those piezoelectric crystals in which the dipoles are permanently lined up even in the absence of an electric field. When such crystals are subjected to an electric field, the direction of polarisation can be changed. This phenomenon of reversal of direction of spontaneous polarisation of a crystal by an electric field is called ferroelectric effect. Its examples are potassium dihydrogen phosphate (KH_2PO_4), potassium tartarate and barium titanate.

FERROMAGNETISM : This is caused by the particles arranged in the lattice in a definite manner and electrons with parallel spins. Iron, cobalt and nickel are ferromagnetic substances. It is a permanent property of the material at a given temperature.

FERROSOFERRIC OXIDE : Fe_3O_4; *See triiron tetraoxide.*

FERROUS COMPOUNDS : Compounds of iron having iron in +2 oxidation state.

FERROUS CHLORIDE : *See iron (II) chloride.*

FERROUS OXIDE : FeO; *See iron (II) oxide.*

FERROUS SULPHATE : $FeSO_4$; *See iron (II) sulphate.*

FERTILISERS : Substances which when added to soil increase its fertility, and subsequently also the crop yield. They should have nitrogen, phosphorous, potassium, calcium, etc. depending on the need of the soil and crop. Urea, ammonium sulphate, ammonium nitrate, potassium chloride, super phosphate of lime etc. are some commonly used fertilisers.

FIBRES : Materials used in textiles, e.g., wool, cotton, asbestos, glass, wool, nylon, polyester, etc.

FILLER : A solid substance used to modify the physical properties and to reduce the price of a synthetic compounds like rubbers, plastics, paints and resins.

Substances which are used as fillers are slate powder, mica, cotton wool, glass, fibre, etc.

FILTER: A device for separating solid particles from a liquid or a gas. The simplest filter may be funnel having a filter paper, or a Buchner funnel with a circular filter paper, or specially made centered glass crucibles which have a porous base as filter. Such crucibles are used in gravimetric exercises and are called Gooch crucibles. The porous filters present in such crucibles are of different porosity and may be used for a precipitate of particular kind.

FILTER PUMP : A simple laboratory suction pump in which air is removed from the filtration flask by a jet of water forced through a narrow nozzle. It causes lowering of pressure inside and causes speedy filtration, particularly in gravimetric exercises. It may be used for vacuum distillation also.

FILTRATE : The clear transparent liquid obtained by filtration.

FILTRATION : The process of separating the suspended solid particles from a liquid using a filter. The substance remaining on the filter is called residue while the substance which passes the filter is called filtrate.

FIVE CHEMICALS : Chemicals of high purity.

FIVE STRUCTURE : Closed spaced spectral lines obtained by the transitions between energy levels which are split up by the rotational or virbational motion of the molecule or the spin of the electron.

FIRE DAMP : Methane gas formed in coal mines.

FIRE EXTINGUISHERS : Materials used to check the spreading of fire, e.g., water, sodium bi-carbonate solution, carbon dioxide foam made from a carbonate and an acid, carbon tetrachloride (pyrene), solids like sodium bicarbonate and potasium bi-carbonate.

FIRST ORDER REACTION : The reaction in which the rate depends only on the molar concentration of one of the species.

$$-d\frac{[A]}{dt} = -k[A]$$

Where k is the specific reaction rate or velocity constant of the reaction. On integration, it leads to -

$$k = \frac{2..303}{t} \log \frac{[A_o]}{[A]}$$

Where A_0 is the active mass of reactant A at zero time and (A) is the active mass of A at time t. Half the period for a first order reaction is given by -

$$t\left(\frac{1}{2}\right) = 0.693 / k$$

Any fractional life period (may be half the period) of a first order reaction remains independent of the initial concentration. Its examples are decomposition of ammonium nitrite and decomposition of hydrogen peroxide.

Hydrolysis of sugar and that of an ester are also the examples of first order pseudounimolecular reactions.

Fischer-Tropsch Process : Process by which synthetic hydrocarbon fuels may be made. In this case, carbon monoxide is hydrogenated in the presence of a catalyst to make hydrocarbons, specially petrol. In fact, water gas is mixed with equal volumes of hydrogen and passed over a heated finely divided nickel for the purpose. The method was extensively used by Germany during World War-II. It is named after the German chemist Frenz Fischer and the Czech, Hans Tropsch.

Fission : (Nuclear fission). The splitting or fragmentation of a large nucleus into smaller nuclei with the release of large amounts of energy. It was first observed by the German scientist Otto Hahn and F. Strauesmann that U-235 splits into two fragments (Kr-94 and Ba-139) when bombarded with slow neutrons. During the process, more neutrons are produced which lead to the chain fission reaction, and so a huge amount of energy is released. The principle is used in making an atom bomb. When carried out in a controlled manner, it may be utilised to produce energy specially electricity. (*See reactors also*).

Fitting Reaction : *See Wurtz reaction.*

Fixation : *See nitrogen fixation.*

Flame : A hot mixture of gases undergoing combustion. The chemical reactions in a flame are generally chain reactions (through free radicals). Light comes from fluourescence of excited molecules or ions or from incandescence of small carbon particles. Luminous part of the zone represents partial combustion of gases while non-luminous zone at the top represents complete combustion. It is the hottest part of the flame.

Flame Retardants : Materials which reduce the tendency to burn and burn by glowing, e.g., borates, phosphates, etc.

Flame Test : A preliminary test in qualitative analysis in which a small amount of sample is burnt in the nonluminous part of the flame of a Bunsen burner, with the help of a clean platinum wire. Some particular colour is observed in the flame which arises due to the emission of light of a definite wavelength by the excited sample in the flame. The principle is applied in spectrographic analysis also. Colours observed in the flame by some of the cations are as follows - barium - grassy green; calcium - brick red; strontium – red; lithium - crimson; sodium - golden yellow; potassium - violet.

Flare Stack : A chimney through which unwanted gases are burnt.

Flash distillation : Rapid distillation or fast removal of the solvent.

Flash Photolysis : A method for studying free radical reactions in gases. The gas is filled in a long glass tube at low pressure. It is then subjected to a very brief intense flash of light from a lamp outside the tube, producing free radicals, which are identified by their spectra. Measurements of change in intensity with time through an oscilloscope help in studying the kinetics of very fast free-radical reactions.

Flash Point : The temperature at which the vapour above a liquid forms a combustible mixture with air and so produces a spark by the contact of naked flame.

Flint : Very hard dense nodules of microcrystalline quartz.

Flocculation : The process of aggregating the smaller particles into bigger ones. It may also be termed as precipitaton or coagulation in case of colloidal solution. Usually the precipitate in this case floats on the surface of the liquid.

Fluidisation : A method of transportation of powders is based on fluidisation in which solid particles suspended in a stream of gas are treated as a liquid. Solid particles suspended in an upward stream are extensively used in the chemical industry particularly in catalytic reactions where the powdered catalyst has a high surface area.

Fluorescein : A yellow-red dye which gives yellow solution with green fluorescence in alkaline medium.

Fluorescence : *See luminescence.*

Fluoridation : The process of adding very small amounts of fluorine salts like sodium fluoride to water to prevent diseases like tooth decay.

Fluoride : A compound having F^- ions.

Fluorination : Process of introducing a fluorine atom into a molecule. *See halogenation also.*

Fluorine : F; A poisonous pale yellow element of group VII of the periodic table. It is the first member of the halogen family with atomic number 9 and relative atomic mass of 18.9984. It occurs as fluorspar (CaF_2) and cryolite (Na_3AlF_6). It is prepared in the laboratory by the electrolysis of molten potassium fluoride and hydrogen fluoride using a copper vessel. Fluorine is the most reactive and most electronegative element. It may react with almost all the elements directly, including xenon, a noble gas. Maximum oxidation state of an element is observed when it reacts with fluorine, e.g., SF_6, PF_5, sulphur and phosphorous exhibit +6 and +5 oxidation states. Unlike other halogens, fluorine does not form oxides or oxyacids. OF_2, Oxygen difluoride on reacting with water gives hydrogen fluoride. Fluorine is highly hazardous and causes severe burns on the skin. The element was identified first by Scheele and first isolated by Moissan about 100 years after its discovery. The cause of this long period was due to its very high reactivity towards the elements of the reaction the vessels used for the preparation of fluorine.

Fluorite : A mineral form of CaF_2. Exists in various colours, most common of which are green and purple. It is usually added as flux material in the smelting of iron and steel. It is also used as a source of fluorine. It finds use in the glass and ceramics industries.

Fluorite Structure : *See calcium fluoride.*

Fluoroboric Acid : HBF_4; Obtained by treating boron triflouride with basic acid. It is a weak acid. BF_4 ions are tetrahedral.

FLUOROCARBON : A compound derived from hydrocarbons by replacing hydrogen atoms with florine atoms. Fluorocarbons are very less reactive and stable even at high temperatures. They are used in aerosol propellents, oils and greases and synthetic polymers such as polytetrafluoroethene. Fluorocarbons are often called freons and their use may cause damage to the ozone layer of atmosphere. Examples are $CHCl_2F$; $CF_2=CF_2$ etc.

FLUORSPAR : A naturally occuring form of calcium fluoride.

FLUX : (a) A substance used during smelting which may combine with impurities to form a fusible mass called slag. For example, lime (from lime stone) acts as a flux for removing acidic impurity of silica in the form of calcium silicate slag. (b) A substance used to keep the metal surface free of oxide in soldering.

FLUXIONAL MOLECULE : A molecule which rearranges so quickly at room temperature that the normal concpet of structure becomes inadequate. In such a molecule, relative positions of atoms cannot be identified and no structure exists for more than 10^{-2} seconds. For example, PF_5 exhibits only one type of fluorine atoms.

FOAM : A dispersion of a gas in a liquid. Foams can be stabilised by surface active agents like soaps and detergents. Foam is made by agitating a liquid in presence of a gas and a stabilising agent. Foams are used in fire fighting. Solid foams, e.g., expanded polystyrene or foam rubber, are made by foaming the liquid and their subsequent setting.

FOLIC ACID : Folacin ; A vitamin of the B-complex category. In its active form, tetrahydrofolic acid, it is a coenzyme in various reactions involved in the metabolism of amino acids, purines and pyrimidines. Green leaves are very rich in this vitamin. It is used in the treatment of sprue and some other types of anaemia. Deficiency causes poor growth and nutritional anaemia.

FOOL'S GOLD : *See pyrites.*

FORMALDEHYDE : HCHO; *See methanal.*

FORMATE : *See methanoate* .

FORMIC ACID : *See methanoic acid.*

FORMULA : A brief representation of a compound with the help of symbols of the various constitutent atoms. The number of atoms may be shown by subscripts. An empirical formula describes the simplest ratio of atoms, while molecular formula gives the types of exact number that are linked to one another in the molecule including their arranegment in space. For example, the empirical formula of acetic acid is CH_2O whereas its molecular formula is $C_2H_4O_2$ and its structural formula given by is CH_3COOH.

FORMULA WEIGHT : The relative molecular mass of a compound obtained from its molecular formula by adding the atomic masses of all the atoms present.

FORMYL GROUP : The group HCO.

FORTIN BAROMETER : An instrument to measure atmospheric pressure.

FOSSIL FUELS : Formed by the decomposition of the remains of living organisms due to high temperature and high pressure below the surface of earth. Important fossil fuels are cola, petroleum and natural gas. Petroleum is not used as a fuel directly but various fractions obtained by the fractional distillation (or refluxing) are used as fuels for different purposes, e.g., petrol, diesel etc.

FRACTIONAL CRYSTALLISATION : Based on the difference in solubilities of solids in a particular solvent, this method is used for separating a mixture of solids. The least soluble substance will crystallise first from the solution of mixture in a hot solvent, on cooling. By controlling the temperature, it is possible to remove the components one by one.

FRACTIONAL DISTILLATION : Separation of a mixture of liquids based on the difference in their boiling points. It can be achieved by distilling the mixture in a flask fitted with a long vertical column called fractionating column. Fractionating columns may be of various types and are made to provide large surface area to the upgoing vapours of liquid components. Various fractions of the mixture can be drawn out at various points of the column depending on temperature. Industrially, fractional distillation is performed in large towers containing perforated trays. Petroleum refining is an example of fractional distillation.

FRANCIUM : Fr; A radioactive element of s-block belonging group IA of the periodic table. It has atomic number 87. It occurs along with uranium and thorium in their ores. Various isotopes are known, of which Fr-223 is the most stable. The element was discovered by Marguerite Perey.

FRASCH PROCESS : A method of drilling out sulphur from earth. The method utilises a tube containing three concentric pipes. Superheated steam is passed down the outer pipe to melt the sulphur which is forced out through the middle pipe. Inner pipe is used for pressing in air under high pressure.

FREE ENERGY : G; A measure of the capacity of a system to do useful work, i.e., work other than the work of expansion. Gibb's free energy is defined by $G = H-T\Delta S$. As such, it is not of much significance, but the change of free energy (ΔG) is useful in indicating the condition for the spontaniety of a process. G is the resultant of two factors responsible for spontaniety – decrease in enthalphy and increase in enthalphy.

$\Delta G = \Delta H - T\Delta S$

IF ΔG is $^{-}$ve , the reaction will be spontaneous, and if $G = O$ the reaction will be at equilbrium.

FREE RADICAL : An atom or group of atoms having an unpaired electron in its valence shell. Free radicals are obtained by the homolytic cleavage of bonds, may be the energy of radiations or heat. Free radicals are highly reactive, and the least stable intermediates in chain reactions like halogenation of alkanes.

FREEZE DRYING : A process used in dehydrating food and other heat sensitive substances. The product is put in deep freezers and the ice formed is sublimed by reducing the pressure.

After the removal of water vapour, the dry product is left behind.

FREEZING : The process of conversion of a liquid into its solid by cooling.

FREEZING MIXTURE : A mixture of two or more compounds which may produce a low temperature, e.g., a mixture of sodium chloride and ice may give a temperature as low as - 20°C.

FREEZING POINT : The temperature at which an equilibrium between a liquid and its solid phase is attained at standard pressure. Below this temperature, liquid solidifies. It is always equal to the melting point of the solid.

FREONS : *See fluorocarbons.*

FREUNDLICH ISOTHERMS : Freundlich gave an empirical equation to account for the extent of adsorption and pressure of the gas keeping the temperatue constant.

Mathematically, it is given by

$$\frac{x}{m} = kp^{1/n}$$

Where k and 1/n are constants to be obtained experimentally. The value of 1/n lies in between 0 and 1. x/m is the amount of gas adsorbed per gram of adsorbent. Logarithmic form of the equation is -

$$\log (x/m) = \log k + 1/n \log p$$

So a graph of log (x/m) versus log p gives a straight line. Slope of the line is equal to 1/n while intercept is equal to log k.

FRIEDEL CRAFT'S REACTION : A type of organic reaction in which an alkyl group or an acyl group may be introduced in the benzene ring by replacing a hydrogen atom. The reactions occur at about 373K in the presence of anhydrous aluminium chloride. The reactions are electrophilic substitutions through the formation of intermediate resonance–stabilised carbocations. Alkenes or alkyl halides may form alkyl benzene while acyl halides form alkylaryl ketones. The reaction is named after the French scientist Charles Friedel and U.S. scientist James H. Craft.

Froth Floatation : A method of concentrating sulphide ores. The well ground mixture is added to water mixed with oil (like eucalyptus or pine oil). Air is blown through to create froth. Ore particles are wetted by the froth and moves upwards while gangue particles remain at the bottom as residue. Froth is shifted to the other part of the vessel and is broken by water shower to get the well concentrated ore. This method is employed to concentrate sulphide ores of copper, zinc, mercury, etc.

Fructose : $C_6H_{12}O_6$; $CH_2OH\text{-}CO\text{-}(CHOH)_3\text{-}CH_2OH$; A simple monosaccharide, stereoisomeric with glucose. It has a ketonic group at C-2 in open chain structure. Further studies reveal the structure to be heterocyclic based on furan ring and so it is called fructofuranose. Fructose, a ketohexose, occurs in green plants, fruits and honey. Naturally occurring fructose has D-configuration and is laevorotatory. It is obtained as a hydrolysis product of sucrose along with glucose.

Fruit Sugar : *See Fructose.*

Fuel : A substance which may produce energy in some useful form (like heat) by oxidation in a furnace or heat engine. For example, wood, vegetable oil, coal, petroleum products (like petrol, kerosene) act as fuel. Nuclear fuels are gaining importance these days to produce electricity.

Fuel Cell : A device to convert the chemical energy of fuel into electrical energy. In this type of cell, fuels like hydrogen gas, methane or carbon monoxide and proper oxidising agents are used as reactants. Simplest of these is the hydrogen - oxygen fuel cell in which hydrogen is oxidised at the anode to form hydrogen ions while oxygen is reduced to water, in presence of H^+ at the cathode. The e.m.f. of this fuel cell is 1.23 V. Fuel cells are cleaner and most efficient sources of energy. They have been used in space vehicles and military equipment.

FUEL OIL : A group of petroleum products (oils) used for the production of heat, e.g., engine oil.

FUGACITY : F; A thermodynamic function used in place of partial pressure in the gaseous reactions. It has the same units as pressure and its value is equal to the pressure in case of an ideal gas. The fugacity of a liquid or solid is the fugacity of the vapour with which it is in equilibrium. The ratio of fugacity to the fugacity in some standard state is called activity. For a gas, the standard state is chosen to be the state at which the fugacity is one. The activity is then equal to fugacity.

FULLER'S EARTH : A clay material which can decolourise oil and grease.

FULMINIC ACID : H-O-C=N; An unstable explosive acid, also called cyanic acid. It is a volatile liquid which readily polymerises. On hydrolysis it gives ammonia and carbon-dioxide. It is isomeric with another acid, H-N=C=O, called isocyanic acid.

FUMARIC ACID : *See butenedioic acid.*

FUNCTIONAL GROUP : An atom or group of atoms on which the properties of a compound depend, e.g., -OH,-CHO,-COOH etc.

FUNDAMENTAL CONSTANTS : The constants which do not change throughout the universe. The gas constant, the Planck's constant, the gravitational constant, the charge on an electron are all fundamental constants.

FUNDAMENTAL UNITS : A set of independently defined units of measurement. In the c.g.s. system the units are centimetre, gram and second for distance, mass and time respectively. In the metric system, they are replaced by metre, kilogram and second (m.k.s. system). This metric system is now further developed to S.I. units in which four more properties are included, i.e., Kelvin for temperature, mole for amount of substance, ampere for electric current and candela for luminous intensity. In British Imperial units, the foot-pound-second (f.p.s.) system was formerly used.

FUNGICIDE : Materials used to prevent fungal attack, e.g., in agriculture, wood, etc. Copper sulphate, 2,4,6–trichlorophenol, Bordeaux mixture (copper sulphate + lime + water) are certain examples of fungicides.

FURAN : C_4H_4O; A colourless liquid, five membered heterocyclic ring. The furan ring contains four carbon atoms and an oxygen atom with two C=C double bonds.

FURANOSE : A sugar having a five membered ring containing four carbon atoms and one oxygen atom.

FURANAL : An aldehyde formed by replacing a hydrogen atom of furan by an aldehyde group. It is a colourless liquid which occurs in essential oils and fuel oils. It is used as a solvent and to test desi ghee.

FUSED RING : See ring.

FUSIBLE ALLOYS : Alloys which melt at low temperature. They are usually eutectic mixtures of bismuth, lead, and cadmium. Wood's metal and Lipòwitz's alloy are the examples.

FUSION : (a) Melting of a solid substance into its liquid phase. (b) Nuclear fusion.

G

GADOLINIUM : Gd; A soft, silvery, ductile metal belonging to lanthanides (4 f series off - block). Its atomic number is 64 and relative atomic mass is 157.25. It occurs in gadolinite, xenotine, monazite etc. It has several stable as well as artificially synthesised isotopes Gd-155 and Gd-157 which are neutron absorbers and so used in nuclear technology. It is also used in alloys, magnets and in the electronics industry. The element was discovered by J.C.G. Marignac.

GALACTOSE : $C_6H_{12}O_6$; A simple monosaccharide, a stereoisomeric glucose, occurs naturally as one of the products of the enzymatic digestion of lactose. Structurally it is an aldohexose.

GALENA : PbS; A mineral form of lead (II) sulphide. It is the chief ore of lead.

GALLIUM : Ga; A soft, silvery, low melting metal belonging to group IIIA (in p-block, B-family) of periodic table. It occurs in minute quantities in ores like that of zinc. Gallite, $CuGaS_2$, is another ore but commericaly, it is extracted from bauxite. Gallium is used in low melting alloys, high temperature thermometers and as adopting element in semiconductros. Gallium arsenide is extensively used as a semiconductors in many devices particularly light emitting diodes. The element was discovered by Francois Lecoq de Boisbaudram.

GALVANISATION : A process of coating steel or iron surface with zinc by diping it in a bath of molten zinc or by the electro–deposition method. It is an effective method of protecting steel from corrosion. Such a protection is at the cost of zinc which is more reactive than iron and so it is called sacrificial protection.

GAMMA RADIATIONS : Electromagnetic radiations emitted by excited nuclei. Gamma waves have very high frequency and short wavelength. A common source of gamma radiations is cobalt-60.

GANGUE : Rocky, other earthen undesirable materials present in an ore.

GARNET : A group of silicate minerals corresponding to the general formula $A_3B_2(SiO_4)_3$. The element A may be magnesium, calcium, manganese or iron (II), while B may be aluminium, iron (III), chromium or titanium. Various garnets are used as gemstones are almandine - $Fe_3Al_2Si_3O_{12}$; spessartite - $Mn_3Al_2Si_3O_{12}$; grossularite - $Ca_3\ Al_2\ Si_3\ O_{12}$; Pyrope - $Mg_3Al_2Si_3O_{12}$.

GAS : A state of matter in which the forces of attraction among the molecules are the weakest and so the particles are in a state of regular random motion. Molecules of the gas are continuously colliding with one another as well as with the walls of the container. Ideal gases obey certain laws called gas laws such as Boyle's law, Charles' law, Graham's law, etc. On the basis of various laws, the properties of gases may be studied.

GAS CHROMATOGRAPHY : A method of separating mixtures of gases of volatile liquids by chromatography. It employs a stationary phase (a solid in case of gas-solid chromatography or a solid coated with a non volatile liquid in case of gas–liquid chromatography) packed in a column. A volatile sample is now introduced and an unreactive carrier gas like nitrogen is passed through. The components of the mixture pass through the column at different rates and are detected as they leave the column.

Two types of detectors may be used; the katharometer which measures changes in thermal conductivity and the flame-ionisation detector which converts the volatile components into ions and denotes the change in electrical conductivity.

GAS CONSTANT : R; The constant that appears in the equation of state for an ideal gas. Since work can be represented in different units, R will have different numerical values in different units–0.0821 L atoms K^{-1} mol^{-1}; 82.1 cm^3atoms K^{-1} mol^{-1}; 8.314 x 10^7 ergs k^{-1} mol^{-1}; 8.314 J K^{-1} mol^{-1} and 1.99 Cal K^{-1} mol^{-1}

GAS EQUATION : pV = nRT, where p = Pressure

V = volume

n = no. of moles

R = universal gas constant

T = temperature

Valid for an ideal gas.

GASEOUS DIFFUSION SEPARATION : A technique based on the principle of diffusion (Graham's) which states that the ratio of diffusion of a gas is inversely proportional to the square root of its density or molecular mass. The isotopes of uranium, U-235 and U-238, may be separated by first preparing UF_6 from uranium ore. It is then passed through a series of fine pores. The separation is attained as the gases pass through successive diaphragms. The method can also be applied to separate isotopes of hydrogen.

GAS LAWS : Laws which relate the temperature pressure, volume and amount of a gas. They are Boyle's law, Charles' law. Avogadro's law and Dalton's law of partial pressures. Boyle's law (V α 1/p at constant T), Charle's law (v α T at constant P) and Avogadro's law (V α n at constant T, P) give rise to universal gas equation the, pV = nRT. All real gases deviate to some extent from the gas laws. These laws are applicable only to ideal gases due to the assumptions that the actual volume of the gas moleucule is negligible and that there are intermolecular forces among the gas, but the most valid of these is Van der Waal's equation.

GAS LIQUID CHROMATOGRAPHY (GLC) : *See chromatography.*

GASOLINE : *See petroleum.*

GAS SOLID CHROMATOGRAPHY (GSC) : *See gas chromatography.*

GAS THERMOMETER : A device for measuring temperature in which the working fluid is a gas. The method is most accurate within the range 25 to 1337 K. Using a fixed mass of a gas, a constant volume thermometer measures the pressure at relevant temperatures usually by mercury manometers.

GATTERMANN'S REACTION : It is the modification of Sandmeyer's reaction for preparing aryl halides from diazonium salts. In this method, a diazonium salt is mixed with freshly precipitated copper powder ($In + CuSO_4$) in the presence of a halogen acid to get the corresponding aryl halide on a little warming. It was discovered by the German chemist Ludwig Gattermann.

GAUCHE CONFORMATION : A conformational isomer of butane obtained by free rotation of C - C bond. The form is partially staggered w.r.t. methyl groups. In this form, the dihedral angle separating two methyl groups is 60° although in the completely staggered form (called anti form) it is 180°.

GAUSS : G ; The unit of magnetic flux density in the c.g.s. system.

GAY - LUSSAC'S LAW : Under similar conditions of temperature and pressure, when gases react together, the volumes of the reacting gases as well as products (if gases) bear a simple whole number ratio. The law is named after the discoverer J.L. Gay Lussac.

Charles' law is also sometimes called Gay Lussac's volume law after the name of its independent discoverer, Gay Lussac.

GAYLUSSITE : A mineral consisting of hydrated mixed carbonate of sodium and calcium, Na_2CO_3. $CaCO_3$. $5H_2O$.

GEIGER - MULLER COUNTER : A device used to detect and measure the radioactivity due to the presence of ionising radiations. It consists of a tube filled with a gas at low pressure (Ar, Ne or CH_4) and fitted with a hollow cathode through the centre of which runs a fine wire anode. A potential difference of about 1000V is maintained between the electrodes. An ionising particle or a photon is allowed to pass through a window into the tube which causes an ion to be produced. Due to high potential difference, the ion will move towards the particular electrode and cause further ionisation by collisions. The consequent current pulses can be counted in electronic circuits. It was first devised by Hans Geiger and finally improved by Geiger and W. Muller.

GEL : A lyophilic colloid that has coagulated to a rigid, jelly-like solid. Gels may be divided into elastic gels (e.g. gelatin) and rigid gels (i.e. silica gel).

GELATIN : A colourless or pale yellow, water soluble protein prepared by boiling collagen with water and evaporating the solution. On the addition of water, it swells, and dissolves in hot water. The solution sets on cooling to make a gel. It is used in photographic emulsions and adhesives, and in jellies and other foodstuff.

GEL FILTERATION : A type of column chromatography in which a mixture of liquids is passed down a column filled with a gel. Small molecules can enter pores in the gel and so move down slowly while the bigger particles pass out rapidly. The method is based on the particle size and is generally used for separating proteins.

GELIGNITE : A highly explosive substance made from nitroglycerine (glyceryl trinitrate), cellulose nitrate, sodium nitrate and wood pulp.

GEM : Molecules in which two functional groups are attached to the same carbon atom. For example, in ethylidene chloride, CH_3CHCl_2, both the chloro groups are on the same carbon atom.

GEOCHEMISTRY : The study of the chemical composition of the earth. It includes the study of the presence of various elements and their isotopes, their distribution in various layers of earth like lithosphere, atmosphere, hydrosphere, etc. and their percentage abundance.

GEOMETRICAL ISOMERISM : *See isomerism.*

GERANIOL : $C_9H_{15}CH_2OH$; A primary alcohol present in a number of essential oils.

GERMANES : Germanium hydrides. Compounds of germanium with hydrogen in which Ge - H bonds are covalent in nature, e.g., GeH_4 and Ge_2 H_6

GERMANIUM : A shiny hard element of group IV of the periodic table having atomic number 32 and relative atomic mass, 72.59. It is present along with zinc sulphide in nature and is obtained as a by–product in the metallurgy of zinc. It is present as argyrodite ($4Ag_2SGeS_2$) also. It was used in early semiconductor devices but has been replaced by silicon now. It is used as an alloying agent, catalyst, phosphor and in infra red equipment. It also forms a large number of organometallic compounds. It's existance was predicted by Mendeleev as eka silicon and was discovered by Winkler.

GERMAN SILVER : An alloy of copper, zinc and nickel. The alloy looks like silver and so is used in cheap jewellery and cutlery.

GIBB'S FREE ENERGY OR GIBBS FUNCTION : G; *See free energy.*

GIBBS-HELMOLTZ EQUATION : G = H T S. The equation is used to predict the spontaniety of chemical reactions proceeding at constant pressure.

GIBBSITE : A mineral form of aluminium hydroxide (hydrated). It is named after a U.S. minerologist George Gibbs.

GIGA : G; A prefix used in the metric system to denote 10^9 times, e.g., 10^9 J (I.G.J.)

GILSONITE : A naturally occurring high melting point variety of bitumen. It has very low water absorption tendency and a high dielectric strength.

GLACIAL ETHANOIC ACID : *See ethanoic acid.*

GLASS : An amorphous, transparent or translucent super cooled solution of various silicates and borates of potassium and calcium. It has no specific formula and does not possess a sharp melting point. Ordinary glass (soda glass) is made by heating sodium carbonate, calcium oxide and silica in a specially designed furnace at about 1700K. Its approximate formula $Na_2SiO_3CaSiO_3SiO_2$. Specific colours may be introduced by adding suitable compounds, e.g., cobalt oxide for blue, cuprous oxide for red, cadmium sulphide for lemon yellow and manganese dioxide for purple colour. Special qualities of glasses are made by changing the composition. For example, hard glass, K_2O CaO. 4 SiO_2 is used in making combustion tubes and chemical apparatus; flint glass, K_2 PbO. $4SiO_2$ has high refractive index and shining lustre and is used for making electrical bulbs, optical instruments and artificial gems Crooks Glass or optical glass contains cerium oxide as one component. which has the capacity to cut off the ultra violet rays which are harmful to the eyes.

Glass Electrode : A type of half-cell having a glass bulb containing a solution of fixed pH into which platinum wires are dipped. The glass bulb is thin enough for hydrogen ions to diffuse through. This bulb is then placed in an unknown solution whose pH is to be determined. The electrode potential depends on the hydrogen ion concentration.

Glass Fibres : Glass drawn into thin fibres some 0.0005 mm to 0.01 mm in diameter. The fibres may be spun into threads and woven into fabrics which are then impregnated with resin to give a material which is very strong and may be used for making boides of cars, boats etc.

Glauberite : A mineral having sodium and calcium, $Na_2SO_4.IOH_2O$; sodium sulphate decahydrate. Named after J.R. Glauber.

Globulin : Any of the globular proteins, insoluble in water, present in blood, eggs, milk, etc.

Glucans : Polymers of glucopyranose.

Gluconic Acid : $CH_2OH(CHOH)_4$ COOH; An optically active acid obtained by the oxidation of aldohexose, glucose. It is soluble in water as well as alcohol. Its calcium salt is used in medicines.

Glucose : $C_6H_{12}O_6$; Dextrose, grape sugar, A white crystalline monosaccharide aldohexose, optically active sugar. It is soluble in water and alcohol. Naturally occuring glucose is dextrorotatory. It is an important compound in the energy metabolism of living organism. It exists in two forms, having specific rotations of +113 and +19 in their freshly prepared aqueous solutions, but the rotation changes with time and at equilibrium, acquires +52.5 value. The concept is called mutarotation. Both these forms exist in pyranose rings as glucopyranose (six membered heterocyclic ring compounds).

Glue : A colloidal mixture of proteins made up of amino acids. It is prepared from waste bones, tendons, skins, etc.

Glutamic Acid : $C_5H_9NO_4$; Aminoglutaric acid. A naturally occuring, dextrorotatory substance present in large amounts in protein hydrolysis products.

Glutaric Acid : $L_5H_9O_4$; Pentanedioic acid.

GLUTEN : A mixture of gliadin and glutelin proteins obtained from wheat.

GLYCERIDES : Esters of glycerol. Acyl groups may appear at one, two or all the three places by replacing hydrogens of hydroxyl groups producing mono, di or triglycerides. Triglycerides of higher carboxylic acids are called oils and fats. If one of these is esterified by phosphoric acid, the compound formed is called phospholipd.

GLYCEROL : $(HO)CH_2$ $CH(OH)$ CH_2 (OH); Glycerine or propane -1,2,3-triol, a trihydric alcohol, colourless, sweet in taste, viscous, highly miscible with water but immiscible with ether. The cause of high viscosity and water solubility is its tendency of undergoing intermolecular hydrogen bonding. It is widely distributed in all the living orangisms, plants as well as animals, in the form of glycerides or oils, fats and lipids. It is obtained as a by–product of soap industry. Usually oils and fats on alkaline hydrolysis (called saponification) produce soaps and glycerol.

GLYCINE : H_2 N CH_2 COOH; Glycocol, the simplest amino acid, sweet in taste and the only amino acid which is optically inactive.

GLYCOGEN : $(C_6H_{10}O_5)n$; Known as animal starch, it is a polysaccharide consisting of highly branched polymer of glucose occuring in animal tissues (particularly liver and muscle cells). It is the main store of carbohydrate energy in the animal cells.

GLYCOL : $HOCH_2$ - CH_2OH; *See ethanediol.*

GLYCOLLIC ACID : CH_2 (OH) COOH; 2-Hydroxy ethanoic acid, a colourless crystalline compound. It occurs in cane juice and beets. It may be obtained in the laboratory by electrolytic reduction of oxalic acid. It is used in textiles and leather industry.

GLYCOLYSIS : A process in which the metabolic break down of carbohydrates takes place.

GLYOXAL : CHO-CHO; Ethanedial. A yellow solid obtained by the oxidation of acetaldehyde or ethylene by selerium oxide. An oxidation product of glycol.

GLYOXYLIC ACID : CHO. COOH; A thick syrupy liquid, widely present in plant and animal tissues. It condenses with urea to form allantoin, an oxidation product of glycol.

GOLTHITE : FeO.OH ; A yellow brown mineral of iron. Maybe formed as a result of oxidation and hydration of iron minerals. Most limonite is composed largely of cryto crystalline golthite.

GOLD : Au; A soft yellow malleable and ductile transition metal of group IB of the periodic table with atomic number 79 and relative atomic mass 196.967. It is present in native state as well as in some lead and copper sulphide ores. Along with silver, it is present as telluride - sylvanite (Ag Au) Te_2. It is unreactive, noble, coinage metal and insoluble in acids, alkalies and other solvents. Gold reacts with aqua regia to form gold (III) chloride. It is used in jewellery (mixed with a little of copper to harden it) dentistry and electronic devices.

GOLD CHLORIDE : $AuCl_3$; A compound obtained by dissolving gold in aqua regia. Exists as a dimer Au_2Cl_6 and is used in photography.

GOLD NUMBER : A measure of the protective power of a lyophilic sol. It is defined as the minimum number of milligrams which when added to 10ml of standard red gold sol (0.0055% gold) just checks its coagulation (change of colour from red to blue) on addition of 1 ml of 10% sodium chloride solution. Larger the gold number, the less efficient it is for protecting lyophobic colloidal solutions.

GOLDSCHMIDT PROCESS : A method of extracting metals by reducing their oxides with powdered aluminum. Metals like chromium, nickel, manganese may be obtained by this method. The principle is also used in thermite welding for heavy, fixed iron structure. It is named after the German scientist, Hans Goldschmidt.

GOOCH CRUCIBLE : A porcelain dish with a perforated base over which a layer of asbestos is placed. It is used for filtration in gravimetric analysis.

GOUY'S BALANCE : A balance used to detect the magnetic nature of a compound, that is, whether or not it is paramagnetic.

GRAHAM'S LAW OF DIFFUSION : The rates at which gases diffuse are inversely proportional to the square roots of their densities (or molecular masses). The law was proposed by Thomas Graham.

Gram - G : Fundamental unit of mass in the c.g.s. units or one thousandth of a kilogram, the unit of mass in S.I. system. Previously, the amount of substance was expressed in gram atomic mass or gram equivalent mass or gram molecular mass, but now these terms are replaced by mole.

Gram Molecular Volume : The volume occupied by one gram molecule (one mole) of an element or a compound in its gaseous state under standard conditions of pressure and temperature. The gram molecular volume of all substances is 22.4 L.

Granite : An igneous rock used as a building material.

Graphite : An allotrope of carbon which is soft, smooth and a good conductor of electricity. It is used as a lubricant, moderator, black lead for pencils and in paints.

Gravimetric Analysis : A type of quantitative analysis based on precipitation of the substance and weighing it in the dry state. For example, barium ions in a given solution can be measured quantitatively by precipitating the same as barium sulphate and weighing the dry precipitate.

Gray : GY; The S.I. unit of absorbed loss of energy per unitmass, resulting from the passage of ionising radiation through a live tissue. One gray is the energy absorbed per kilogram of mass.

Greases : Solids or semifluids prepared by mixing a thickening agent in a liquid lubricant (oil). Resistants to temperature, greases may contain soaps of sodium, calcium, aluminium etc. Non-soap greases contain organo clays or some dyestuffs.

Greenockile : A mineral form of cadmium (II) sulphide.

Green Vitriol : $FeSO_4$. *See iron (III) sulphate.*

Grignard's Reagent : A class of organometallic compounds having magnesium metal. Represented by the general formula. RMgX, where R is an organic group while X is a halogen atom, e.g., C_2H_5MgBr, CH_3MgI, etc. Their structure is supposed to be R_2 Mg. Mg X_2. They may be made by refluxing an alkyl halide with magnesium in dry ether. They are highly unstable, explosive compounds and so are not isolated. Their ether solution is used as such in reactions. Grignard's reagents are very useful compounds to synthesise a very large number of organic compounds. With water, alcohol or an amine they form alkanes, with methanal they form primary alcohol with other aldehydes they form secondary alcohols and with ketones they form tertiary alcohols.

The compounds are named after the discoverer, the French chemist F.A.V. Grignard.

GROTTHUS-DRAPER LAW : The law dealing with photochemical changes. When light falls on a body, a part of it is reflected, a part of it is transmitted and the rest is absorbed. It is only the absorbed light which is effective in bringing about a chemical process.

GROUND STATE : The lowest stable energy state of a system such as a molecule or atom or ion.

GROUP : A group in the periodic table represents a series of chemically similar elements which have similar electronic configuration and have a gradation in their physical properites. There are 18 groups or vertical columns in the long form of periodic table.

GUANIDINE : A crystalline basic compound $HN{=}C(NH_2)_2$, related to urea.

GUANINE : A purine derivative, an important component base of nucleotides and the nucleic acids DNA and RNA.

GUN COTTON : *See cellulose nitrate.*

GUN METAL : An alloy of copper and tin having 10% tin, used to make gun barrels, castings and bearings.

GUN POWDER : A powdered explosive made by mixing sulphur, charcoal and potassium nitrate.

GUTTA PERCHA : A naturally occuring polymer which is used as a rubber additive obtained from tropical trees.

GYPSUM : $CaSO_4.2H_2O$; A mineral form of calcium sulphite used to manufacture plaster of paris.

H

HABER'S PROCESS : A method of manufacturing ammonia by the reaction of hydrogen and nitrogen under suitable conditions of temperature and pressure, i.e., about 723 K temperature and 250 atmospheres pressure.

Ex : $N_2 + 3H_2 \rightarrow 2NH_3$

The reaction is reversible and exothermic (in the forward direction). The conditions mentioned above are given by the application of Le Chatelier's principle.

This process ways developed by the German chemist Fritz Haber and was industrially developed by Carl Bosch. So it is also called Haber Bosch process.

HAEM : (Heme) A complex compound containing an iron atom which binds with proteins as a cofactor or prosthetic group to form haemoproteins, e.g., haemoglobin, myoglobin, etc.

HAEMETITE : A mineral form of iron (II) oxide, Fe_2O_3. It is a principal ore of iron and is responsible for the red colour of rocks. In industry, it is used as a polishing agent and also in paints.

HAEMOGLOBIN : A globular protein present in the blood of animals which serves as an oxygen carrier. In it, a basic protein called globin is linked with four haem groups. It is capable of combining reversibly with one molecule of oxygen per iron atom of haem groups to form oxyhaemoglobin (a bright red coloured compound). Iron is present in a divalent state. Dissociation occurs in the tissues having low pressure of oxygen to form haemoglobin again. Some other compounds are found to combine with it. For example, carbon monoxide binds more strongly than oxygen and thus behaves as a poison.

HAFNIUM : Hf ; A silvery metal of the 6th period and 4th group having atomic number 72 and relative atomic mass 178.49. Being a transition element, it resembles zirconium (due to lanthanide contraction) and occurs also with zirconium. It exists in the form of compounds having +3 (less stable) and +5 oxidation states. It is used in tungsten alloys in filaments and electrodes. In the reactors, it may be used as a neutron absorber. The element was discovered by Urbain and later on confirmed by D.Coster and G.C. de Hevesey.

HALNIUM : Ha; A transition element of the 7th period having atomic number 105. It is formed only artificially.

HALF CELL : An electrode in contact with suitable ions in a cell. Simplest of these is metal in contact with its own ions in solution, e.g., zinc plate dipped in zinc (II) sulphate solution. Gas half cells have a platinum plate in a solution with gas bubbled over the plate, e.g., hydrogen electrode. Other type of half cells may be a calomel electrode which uses an insoluble salt alongwith a solution in calomel half cell (or electrode) mercury (I) electrode is used with potasium chloride solution.

HALF LIFE ($T_{1/2}$) : The time in which the initial concentration of a reactant is reduced to its one half. The term is commonly used in the kinetics of ordinary chemical reactions as well as in the kinetics of radioactive substances. For first order reactions, it is independent of the initial concentration of the reactant. In radioactivity, it is an important property by which the isotopes are described.

HALF-THICKNESS : The thickness of a particular matter which may reduce the intensity of a beam of radiation to its one half.

HALIDE : A compound of any halogen with some other element of another group. Halides of metals are generally ionic (e.g., NaF, KCl, etc.) but may be covalent also (e.g., $AlCl_3$, BeF_2 etc.). Some classes of organic compounds include alkylhalide (RX) and acylhalide (RCOX). etc.

HALITE : (Rock Salt) ; Naturally occuring sodium chloride. It is colourless or white but may acquire some colour in the presence of impurities.

HALL - HEROCULT CELL : An electrolytic cell used for the extraction of aluminium from bauxite. Pure Al_2O_3 is obtained from bauxite by leaching with caustic soda solution. Pure oxide is then mixed with cryolite. Al_2O_3 is not a good conductor as it melts at a high temperature. Cryolite lowers the melting point as well as converts aluminium of oxide to the aluminium of cryolite which is ultimately obtained at cathode (graphite lining of the cell) and is tapped off. Oxygen is set free at the graphite anode and it gradually corrodes the anode. The cell is named after the U.S. chemist Charles Hall and the French chemist Paul Heroult, who independently discovered the method of extraction.

HALOLKANES : (Alkylhalides) Organic compounds obtained by the substitution of one or more hydrogen atoms of alkane by halogen atoms.

HALOFORM REACTION : Products prepared by the direct reaction between alkane and halogen or by the addition of halogen and/or phosphorus halide or an alcohol. Examples are : dichloromethane CH_2Cl_2, dibromomethane CH_2Br_2 chloroform $CHCl_3$. They may be prepared by the action of bleaching powder on propanone or ethanol (which first changes into ethanal). Iodoform can be made by using iodine and caustic soda (or sodium carbonate) in place of bleaching powder.

$$CH_3\ CO\ CH_3 + 3Na\ Oi\ CH_3\ CO\ Cl_3 + 3\ NaOH$$
$$CH_3\ CO\ Cl_3 + NaOHCH_3COONa + CHI_3$$

As iodoform is a yellow, insoluble solid, the reaction is used to identify methyl ketones and ethanal and the compounds which produce them under given conditions.

HALOFORMS : Compounds with the general formula CHX_3, where X is a halogen atom. Chloroform, bromoform, iodoform are common examples.

HALOGENATION : A chemical reaction in which a halogen atom is introduced into a compound. May be by substitution oı addition. Alkanes are halogenated in presence of *nv* or high temperature to form substituted halogen derivatives like chloromethane, dichloromethane etc. The reaction proceeds through free radicals. Aromatic hydrocarbons undergo halogenation in the dark in the presence of some halogen carrier like ferric halide or iron or aluminium chloride (e.g., formulation of chlorobenzene). The reaction is an electrophilic substitution reaction. In presence of benzene, it may form an addition compound (e.g, $B_1H_1C_1$.) Alkenes and alkynes undergo addition reactions with halogens easily.

HALOGENS : Elements of group VII A of the periodic table–fluorine, F; chlorine, Cl; bromine Br; iodine I; astatine, At. They have similar electronic configuration (ns^2 np^5) in the valence shell. They are highly electronegative elements. Their electron affinity and electonegativity values are maximum in their respective periods. So they have maximum tendency to gain electrons and act as strong oxidising agents. They are typical non-metals and may form covalent compounds with other non-metals while they form ionic compounds with metals. All of them exist as dimers. Their melting points and boiling points increase down the group. Fluorine and chlorine are gases at room temperature, bromine is a liquid–while iodine and astatine are solids. They are highly reactive elements. Within the group, reactivity decreases down the group. Fluorine is most electronegative and shows -1 oxidation state while others may show positive oxidation states (+1 to +7) in their compounds. The oxidising power of halogens is in the order $F_2 > Cl_2 > Br_2 > 1_2$ and the reducing power of their ions is ins the order $I > Br > Cl > F$. With hydrogen they form hydrogen halides or halogen acids. Strength of halogen acids is in the order $HI > HBr > HCl > HF$. They also form oxyacids such as $HOCl$, $HClO_2$, $HClO_3$ and $HClO_4$. Halogens react with one another also tn form inter halogen compounds, e.g., ICl, ICl_3 IF_7 etc.

HAMMER MILLS : Devices used in the industry for crushing and grinding solid materials. By the impact of grinding, particles of suitable specific size may be obtained.

HAMMICK-ILLINGWORTH RULE : An empirical rule describing the directive influence of a group already present on the benzene ring. To illustrate consider C_6H_5 - XY as the compound (Y is attached to X while X to the ring). (a) Y is in a higher group in the periodic table or in the same group as X but of lower atomic weight; then they group XY is meta directing. (b) Y is in the lower group than X or XY is just a single atom ; then the group XY is ortho/ para directing.

Groups like - Cl, -Br, -OH, $-CH_3$ are ortho/ para directing and $- NO_2$, -COOH, -CHO etc are meta directing according to the above rule.

HARDENING OF OILS : The process of passing hydrogen through a heated oil in the presence of a catalyst. By this unsaturated esters (oils) are converted to saturated esters (fats) used to convert vegetable oils to vegetable ghee.

HARDNESS OF WATER : The presence of soluble salts of calcium and magnesium in water. Hardness causes problems in washing, reduces the efficiency of boilers and other industrial processes. Hard water forms a scum with soap and checks lather formation. Temporary hardness is due to dissolved bicarbonates and may be removed by boiling.

Calcium carbonate formed by the decomposition of bicarbonate forms scales in kettles, boilers, pipes etc. Permanent hardness is due to chlorides and sulphates calcium and magnesium. It can be removed by treating with sodium carbonate and the treated water may be used for washing. In order to get drinking water, permutite or some other ion exchange resins may be used. Calgon is another compound by which potable water is obtained. Hardness of water can be estimated by titrating it against EDTA solution of known strength.

HEAT : Energy transferred as a result of temperature difference. It is generally used for the internal energy of the system. It may further be defined in terms of heat of combusion, neutralisation, solution, hydration, vaporisation, fusion etc. which are enthalpy changes for particular processes for 1 mole of the compound. If the substances involved are in their standard state, i.e., 298 K temperature and 1 atmosphere pressure, the heat changes are termed as standard enthalpy of reaction, formation, combustion etc.

HEAT CAPACITY : Thermal capacity. The amount of heat supplied to an object to raise its temperature by 1°C. The molar heat capacity is the amount of heat supplied to raise temperature 1°C of one mole of compound. It is measured in JK^{-1} mol^{-1}. Heat capacity may be measured either at constant volume (C_V) or at constant pressure (C_P). Specific heat capacity is used for heat required for unit mass of the substance to raise temperature its by 1°C.

HEAT ENGINE : A thermodynamic device for converting heat energy into work. It operates between a high temperature source and a low temperature sink. Efficiency of a heat engine depends on the temperatures of source and sink and not on the materials involved.

HEAT OF ATOMISATION : The amount of heat required to dissociate one mole of a given compound into its constitutent atoms.

HEAT OF COMBUSTION : The amount of energy liberated when one mole of a given substance is completely oxidised in excess of air or oxygen.

HEAT OF CRYSTALLISATION : The amount of heat released when one mole of any substance crystallises from its saturated solution.

HEAT OF DISSOCIATION : The amount of heat required to dissociate one mole of a substance into its constituent elements.

HEAT OF FORMATION : The amount of energy released or absorbed when a mole of compound is formed from its constituent elements.

HEAT OF NEUTRALISATION : The amount of heat released when one gram equivalent of an acid, or a base is completely neutralised by the base or an acid respectively, in dilute solutions.

Heat Of Reaction : The amount of heat absorbed or released as result of the complete chemical reaction of molar quantities of the reactants.

Heat Of Solution : The amount of heat energy absorbed or released when one mole of a given substance is completely dissolved in an excess of solvent so that no further heat change takes place on dilution.

Heavy Hydrogen : See deuterium.

Heavy Spar : A mineral form of barium sulphate.

Heavy Water : Deuterium oxide, D_2O; Water in which hydrogen atoms are replaced by deuterium, ${}^2_1H$, atoms. It is a colourless liquid occurring in ordinary water in very small amounts (0.003%). It is prepared by the electrolysis or fractional distillation of water. It is used in nuclear reactors as moderators to slow down fast moving neutrons.

Hecto : h ; A prefix denoting 100 times, e.g., 100 m = 1 hectometre or 1 hm.

Heisenberg's Uncertainty Principle : It is impossible to measure accurately and simulaneously the position and momentum of a fast moving microscopic particle like an electron. It arises because, in order to detect the particle, radiation has to be reflected and it disrupts the position of the particle.

It may be explained in terms of wave particle duality of the particles and radiations. It has no significance in actuality, as the particles being observed are very large in size and remain unaffected by the impact of radiation.

Helium : He; A colourless, odourless, non metal, inert element of the zero group of the periodic table having atomic number 2 and relative atomic mass 4.0026. It has the lowest boiling point of all substances. It exists mainly as He-4 with a small amount of He-3. There are some short-lived isotopes also, like He-5 and He-6. It occurs in the ores of uranium and thorium. It is used to provide inert atomsphere for welding and semiconductor manufacture, as a refrigerant for super conductors and in filling ballons. It is used by deep sea divers and asthmatic patients in the form of oxygen-helium breathing mixture.

Helium is unusual in the fact that it is the only known substance for which there is no triple point (a point at which all the three phases may exist together). At 2.2 K, helium undergoes a trasition from liquid He-I to liquid He-II, the latter being a true liquid but exhibiting super conductivity and extremely low viscosity. The low viscosity allows the liquid to spread in layers, a few atoms thick, described as liquid flowing uphill. Helium was discovered in the solar spectrum by Lockyer.

HELL-VOLHARD-ZELINSKY REACTION : A method for preparing halogen substituted carboxylic acids. When a carboxylic acid is reacted with free halogen in the presence of phosphorous or its halide, halogen gets substituted in place of an alpha hydrogen atom. For example, acetic acid may be converted to mono, di and trichloroacetic acids by chlorine in the presence of phosphorous.

HELMOLTZ FREE ENERGY (FUNCTION) : F; A thermodynamic function defined as $F = E - TS$, where E is the internal energy, T is the temperature and S in the entropy. It is a measure of useful work obtainable from a system working isothermally.

HEMIACETALS : *See acetals.*

HEMIHYDRATE : A crystalline hydrated compound having one water molecule of crystallisation per two molecules of the compound, e.g., $2CaSO_4.H_2O$.

HEMIKETALS : *See ketals.*

HEMIMORPHITE : $Zn_4\ Si_2O_7\ (OH)_2\ H_2O$; An ore of zinc.

HENDERSON : HASSELBACH EQUATION : An equation used to calculate pH of buffer solutions. For an acidic buffer(a mixture of a weak acid and its salt with a strong base) pH is given by

$$pH = pk_a + \log \frac{Salt}{Acid}.$$

Where k_a is the dissociation constant of acid and pk_a is the negative logaritham of k_a.

HENRY : H ; The S.I. unit of inductance. It is equal to the inductance of a closed circuit in which an e.m.f. of one volt is produced when the current in the circuit varies uniformly at a rate of one ampere per second.

$H = 1\ Wba^{-1}$ Wb represents Webre (unit of magnetic flux).

HENRY'S LAW : The law describing the relation between the concentration of gas dissolved to the partial pressure of that gas in equilibrium with solution. Mathematically, $p=K[C]$ where K is proportionality constant. From this, it may be derived that volume of the gas dissolved is constant and remains independent of the pressure.

HEPARIN : An organic compound obtained from the liver which prevents blood clotting and so may be used to treat thrombosis.

HEPTAHYDRATE : Compound having seven molecules of water of crystallisation.

HEPTANE : $CH_3\text{-}(CH_2)_5\text{-}CH_3$; A straight chain, saturated hydrocarbon, colourless liquid used to determine octane number of a fuel. It is assigned an octane number zero.

HEPTAVALENT : Having a valency of seven.

HEROIN : $C_{21}H_{23}NO_5$; A morphine salt, used in the form of its hydrochloride which is soluble in water and alcohol. It is toxic and addictive in nature.

HERTZ : H_Z; The S.I. unit of frequency, equal to the number of vibrations per second. Named after the German scientist, Heinrich Hertz.

HESS'S LAW : The overall energy change when reactants are converted to products remains same whether the reaction proceeds directly in one step or through several steps. It is always independent of the route taken.

The law is a direct consequence of first law of thermodynamics. The law is used to obtain thermodynamic data that cannot be directly and experimentally determined. With its help, enthalpies of transition (from one allotropic form to the other), enthalpies of formation and enthalpies of phase changes can be calculated easily by the addition and subtraction of thermochemical equations from one another. The law is also called law of constant heat summation. It was discovered by the Russian chemist G.H. Hess.

Hetroatom : An atom which differs from others in a ring compound, e.g., oxygen is a heteroatom in furan, S is heteroatom in thiophene and N is a heteroatom in pyrrole.

Heterocyclic : *See cyclic compounds.*

Heterogenous Catalysis : *See catalyst.*

Heterolytic Fission : The breaking of a bond to produce two oppositely charged ions, eg. HCl breaking into H^+ and Cl^- ions.

Hexachlorobenzene : C_6Cl_6; A colourless, crystalline solid prepared from benzene and used as a wood preservative.

Hexadecane : Cetane; $CH_3(CH_2)_{14}CH_3$; A colourless liquid, straight chain alkane used in assigning cetane number to various diesel fuel. It is assigned a standard value of 100.

Hexadecanoic Acid : *See palmitic acid.*

Hexagonal Close Packing : *See close packing.*

Hexagonal Crystal : *See crystal system.*

Hexamine : Hexamethylene tetramine; Urotropine; $(CH_2)_6 N_4$; A white crystalline compound obtained from ammonia and formaldehyde. Used as an antiseptic for urinary diseases.

Hexane : $CH_3(CH_2)_4CH_3$; A colourless liquid straight chain alkane, used as a solvent.

Hexanedioic Acid : $HOOC(CH_2)_4COOH$; Adipic acid; A dicarboxylic acid used in the manufacture of nylon-66.

Hexanoic Acid : $CH_3(CH_2)_4COOH$; Caproic acid. Occurs in the form of its glycerides in cow and goat milk as well as in some vegetable oils.

Hexose : A monosaccharide (carbohydrate) having six carbon atoms in its molecule, e.g., glucose, fructose etc.

High Speed Steel : An alloy of iron having 12-22% tungsten, about 5% chromium and 0.4 - 0.7% carbon. Remains hard even at red heat.

Hippuric Acid : $Ca\ H_9\ O_3\ N$; Benzoylglycine Present in the urine of mammals.

HISTIDINE : *See amino acids.*

HISTOCHEMISTRY : The study of the distribution of the chemical constitutents of tissues by means of their reactions.

HOFMANN'S DEGRADATION : A method of preparing primary amines from acid amides. The reaction is called degradation because one carbon atom is decreased during the process. The reaction involves the refluxing of amide with aqueous caustic alkali and bromine.

$RCONH_2 + 4KOH + Br_2$ -

The method can be applied to prepare aromatic amines also.

HOFMANN'S METHOD : (a) An old outdated mehtod to determine vapour density of a volatile liquid. Advantage of this method is its application to volatile liquids which decompose at their boiling points.

(b) Process of reacting an alkyl halide with ammonia in a sealed lube to get mixture of primary, secondary and tertiary amines as well as quaternary ammonium salt.

HOLMIUM : Ho; A soft, malleable, silvery metal of the lanthanide series having atomic number 67 and relative atomic mass 164.93. It occurs with other lanthanides particularly apatite and xenotite. It was discovered by P.T. Cleve and J.L. Soret.

HOMOCYCLIC : A ring made up of some atoms, e.g., cyclohexane, benzene.

HOMOGENOUS : Describing single phase. A homogenous mixture has got only only one phase.

HOMOLOGOUS SERIES : A group of structurally similar organic compounds arranged in the order of increasing molecular mass. Two adjacent members differe from each other by one carbon and two hydrogen atoms. Members may be represented by a general formula, possess similar chemical properties and may be prepared by a set of similar methods called general methods of preparation. Members of a series are called homologues. For example, the homologous series of alkyl halides CH_3Cl, C_2H_5 ClC_3H_7, etc. has general formula $C_nH_{2n+1}Cl$

HOMOLOGUES : *See homologous series.*

HOMOLYTIC FISSION : (Homolysis); The breaking of a bond in a molecule in which the fragments are free radicals. Chlorine breaks to give free radicals by heat or ultra violet radiations.

HOMOPOLYMER : *See polymer*

HORMONE : A substance made in very small amounts by the endocrine glands of an organism which regulates the growth and functioning of a particular organ in a distant part of the body. For example, insulin regulates glucose metabolism in the body.

HORNBLENDE : A group of commonly occuring rock forming minerals of the amphibole group consisting mainly of the silicates of calcium, iron and magnesium.

HUCKEL'S RULE : *See aromatic compounds.*

HAMECTANT : A substance used to preserve the moisture content of the matter, e.g., glycerol.

HUMIDITY : Absolute humidity is the mass of water vapour per unit mass of air. The relative humidity is the ratio of the partial pressure of the water vapour in the air to the partial pressure of the water vapour in air saturated at that temperature.

HUND'S RULE : A rule that describes the filling of electrons in degenerate orbitals of a particular sub energy level to have minimum pairing. Pairing takes place only when each orbital of a particular subshell has got an unpaired electron. Further the spins of all these unpaired electrons of a subshell are parallel.

HYBRID ORBITAL : *See orbital.*

HYBRIDISATION : A hypothetical process in which different orbitals of nearly the same energy may combine to form an equal number of new orbitals having identical characters and same energy. The new orbitals are called hybrid orbital. Types of hybridisation and based on the number and type of orbitals combining. Geometry of the molecule is based on the type of hybridisation. For example, sp^3 hybridisation leads to tetrahadral, while d^2sp^3 and sp^3d^2 hybridisations lead to the octahedral geometry.

HYDRACIDS : Protonic acids like HF, HCl, HBr, HI.

Hydrate : A substance having some molecules of water of crystallisation, e.g., $CaSO_4.2H_2O$; $CuSO_4. 5H_2O$, etc.

Hydrated Alumina : *See aluminium hydroxide.*

Hydration : Conversion of an anhydrons salt to a particular hydrated state having a definite number of water molecules of crystallisation.

Hydrazine : $NH_2 - NH_2$; A colourless liquid, soluble in water as well as alcohol. It is preapred by reacting ammonia with sodium hypochlorite (Rarchig's method). It is a weak diacidic base giving two series of salts. It is powerful reducing agent and reacts vigorously with oxidising agents so it is used as a rocket propellent.

Hydrozoic Acid : *See hydrogen azide.*

Hydrozones : Organic compounds formed by the nucleophilic addition of hydrazine on carbonyl compounds and subsequently losing a molecule of water. For example, ethanal forms ethanal hydrazone, $CH_3CH = N\text{-}NH_2$.

Hydrobromic Acid : *See hydrogen bromide.*

Hydrocarbons : Compounds containing only carbon and hydrogen. They may be alkanes, alkenes, alkynes, arenes, cycloalkanes, etc. They are the parents of all organic compounds. They occur in nature in petroleum and in coal tar.

Hydrochloric Acid : *See hydrogen chloride.*

Hydrocyanic Acid : *See hydrogen cyanide.*

Hydrofluoric Acid : *See hydrogen fluoride.*

Hydrogen : H; A colourless, odourless, lightest, gaseous element of the first group of periodic table having atomic number 1 and relative atomic mass 1.008. It exists in three isotopes naturally occuring, H-1, and H-2, radioactive, H-3. Hydrogen gas is diatomic and has two forms, ortho hydrogen in which the nuclear spins are parallel, and para hydrogen in which the spins are opposite. At normal temperature it contains 25% para hydrogen while in liquid hydrogen 99.8% is in the para form. It can be obtained by electrolysis of water. It is used in various industiral processes like manufacture of ammonia, production of petroleum from coal, hydrogenation of vegetable oils. Chemically, hydrogen reacts with a large number of metals to form hydrides and non metals to form hydracids and other compounds. It is combustible and catches fire violently. It has very high calorific value but can not be used as a domestic fuel due to high risk involved in its storage and use. It is used in fuel cells to make electricity particularly for space crafts. It was discovered by Henry Cavendish.

Hydrogenation : Addition of hydrogen to an unsaturated compound. Reaction takes place in presence of nickel, palladium or platinum.

Hydrogen Azide : (Hydrazoic Acid); HN_3; A colourless liquid, highly poisonous, strong reducing agent which causes explosion in the presence of oxygen and other oxidising agents. It is prepared by distilling a mixture of sodium azide (NaN_3) and a dilute acid. Its salt lead azide, $Pb(N_3)_2$, is used in detonators because of its ability to explode.

Hydrogen Bond : An electrostatic force of attraction between the highly electronegative atom of one molecule or part of the molecule and hydrogen atom of another molecule or part of the molecule. It is exhibited by compounds having electronegative atoms like F, N and O. The bond is weaker than those of ionic and covalent, but stronger than that of Waal's forces of attraction. H-bonds which join several molecules together (causing association) are called intermolecular H-bonds while the hydrogen bonds which link two parts of the same molecule are called intramolecular H-bonds.

Unusual properties of water like high boiling point, higher density of water than ice, maximum density of water at 277K can all be explained by intermolecular H-bonding. Solubility of a large number of organic compounds is also due to this property. H-bonding occuring within the chains of DNA and between C=O and N-H groups in proteins are its example in living organisims.

Hydrogen Bromide : HBr; A colourless gas made directly by the action of hydrogen and bromine in presence of platinum catalyst. Its aqueous solution is a stronger acid than hydrochloric acid.

HYDROGEN CARBONATE : HCO_3; Bicarbonate.

HYDROGEN CHLORIDE : HCl; A colourless fuming gas. It can be made by the action of chloride salt with concentrated sulphuric acid. Previously known as spirit of salt or Muriatic acid. Industrially, it is prepared from elements. It is a strong acid which ionises completely in solution. It is used in the manufacture of a large number of chloro compounds, PVC, as a laboratory reagent, etc.

HYDROGEN CYANIDE : HCOR; Also called hydroxyanic acid or Prussic acid. It is a colourless gas with a chraracteristic odour of almonds. It is formed by the action of dilute acids on metal cyanides or by the catalytic oxidation of ammonia and methane with air. It is an extremely poisonous compound used in producing acrylate polymers.

HYDROGEN ELECTRODE : (hydrogen half cell); A type of half cell in which hydrogen gas in bubbled over a platinum coil immersed in a solution of hydrogen ions. If the concentration of H^+ ions is 1M, the gas is passed at 1 atmosphere and the temperature is 298K, it is called normal hydrogen electrode or standard hydrogen electrode. Its value is arbitrarily fixed at zero volt. Standard reduction potentials of other cells are measured with its help. It is represented as -

$Pt\ (s)\ /H_2(g),\ H^+{}_{(ar)}$

HYDROGEN FLUORIDE : HF; A colourless liquid made by the action of sulphuric acid on calcium fluoride. The compound is extremely corrosive. It is a fluorinating agent which may attack glass. Due to the small size of fluorine and very high electronegativity, it shows hydrogen bonding resulting in its high boiling point. The bond between hydrogen and fluorine is strong and short, which makes the acid weak.

HYDROGEN IODIDE : HI; A colourless gas made by direct combination of the elements. It is the strongest of all halogen acids. It acts as a strong reducing agent also.

HYDROGEN PEROXIDE : H_2O_2; A colourless unstable liquid. It is prepared by the action of cold dilute sulphuric acid on sodium peroxide or barium peroxide. It may be prepared on a large scale by the electrolysis of sulphuric acid or ammonium hydrogen sulphate. It decomposes in the presence of metal ions or in light to give water and oxygen. Hydrogen peroxide acts as an acid, an oxidant and as a reductant. It is usually represented by its volume strength. For example, 10 volumes of hydrogen peroxide means that its one volume would yield 10 volumes of oxygen on decomposition. Due to its oxidising nature, it acts as an antiseptic as well as a bleaching agent for hair, cloth, feathers, etc.

HYDROGEN SPECTRUM : Spectrum obtained by passing an electric discharge through hydrogen gas taken in the discharge tube at low pressure. The emitted light is analysed with the help of a spectroscope. The spectrum consists of a large number of lines appearing in different regions of wave lengths. J.J. Balmer developed a relationship among the different wavelengths of the series of visible lines :

$$\bar{\mu} = \frac{1}{\gamma} = R\left(\frac{1}{2^2} - \frac{1}{n^2}\right)$$

Where n may be 3,4,5,

But it was valid for spectral lines in the visible region. The relationship was further developed by Rydberg as :

$$\bar{\mu} = \frac{1}{\gamma} = R\left(\frac{1}{n_1^2} - \frac{1}{n_1^2}\right)$$

Where n_1 is fixed for a particular series while n2n1 and may vary. Value of R is 109678 cm^{-1}

Different series of hydrogen spectrum are named as Lyman (U.V. region), Balmer (visible region), Paschen, Brackett and Pfund (all these three in I.R. region).

HYDROGEN SULPHATE : (Bisulphate); A salt of $4SO_4$ ion or an ester of type $RHSO_4$ where R is an alkly group.

HYDROGEN SULPHIDE : H_2S; A colourless gas with the smell like rotten eggs. It is soluble in water and ethanol. It is prepared in the laboratory by the action of dilute sulphuric acid on iron (II) sulphide in Kipp's apparatus which gives intermittent supply. It is an acidic compound having reducing properties. The compound is an important laboratory reagent for the analysis of metals of group II and group IV of qualitative analysis scheme. It is a poisonous compound. It is found in volcanic emissions also.

HYDROGEN SULPHITE : (Bisulphite); HSO_3 ; A salt containing HSO_3 ion or an ester of the type $RHSO_3$ where group R is an alkyl group.

HYDROODIC ACID : *See hydrogen iodide.*

HYDROLYSIS : Reaction between a compound and water. Salts having either weak acid or weak base as a component hydrolyse in aqueons solution resulting in the acidic or alkaline character of its aqueons solution. For example, $CuSO_4$ is hydrolysed to give an acidic solution.

$$Cu^{++} + SO_4 - + 2H_2O = Cu(OH)_2 + 2H^+ + SO_4^{--}$$

HYDROMAGNESITE : A mineral form of basic magnesium carbonate, 3 $MgCO_3$, $Mg(OH)_2.3H_2O$.

HYDROPHILLIC : Compounds that are attracted to water.

HYDROPHOBIC : Having hatred towards water.

HYDROQUINONE : Benzene - 1,4 - diol.

HYDROSOL : (aquosol); A sol or colloidal solution in which dispersion medium is water.

HYDROSULPHATE : HSO_4; Hydrogen sulphate.

HYDROSULPHIDE : HS; Compounds having HS ions like sodium hydrosulphide NaHS. Such compounds are very useful in the preparation of alcohols.

HYDROSULPHURIC ACID : Solution of hydrogen sulphide gas in water. It contains HS and 5- ions along with a small amount of H+ ions due to which solution is weakly acidic.

HYDROXIDE: A compound containing OH^- ion or -OH group. Hydroxides of metals are basic, those of metalloids are amphoteric and these of non metals are acidic in nature.

HYDROXYCERRUSILE : *See lead (II) carbonate hydroxide*

2–HYDROXY PROPANOIC ACID : *See lactic acid*

HYGROSCOPIC : Describing a compound which absorbs moisture from the atmosphere.

HYPERCONJUGATION : The amount of the effect in which alkyl groups interact electronically with the II electron cloud of a double bond of carbon chain. It may explain the stability of highly branched alkenes.

HYPERONS : Elementary particles present cosmic rays. Mass of each particle is 2185 times that of an electron. They are high energy particles.

HYPERTONIC SOLUTION : A solution which has a higher osmotic pressure as compared to the given solution.

HYPO : (Sodium thiosulphate); $Na_2S_2O_3$; A compound used as a fixing agent in photography.

HYPOCHLORITE : M(ClO); A salt of hypocnlorous acid.

HYPOCHLOROUS ACID : HOCl; Used as a bleaching agent and oxidising agent.

HYPOPHOSPHOROUS ACID : *See phosphoiric acid.*

HYPOSULPHITE : *See sulphinate.*

HYPOSULPHUROUS ACID : *See sulphuric acid.*

HYPOTONIC SOLUTION : A solution which has a lower osmotic pressure than the other given solution.

I

ICE : Water in the solid phase. *See water for details.*

ICE POINT : Freezing point of water under standard conditions, i.e., the temperature at which water and ice are in equilibrium at one atmosphere pressure.

IDEAL CRYSTAL : A crystal with no imperfections, impurities or defects.

IDEAL GAS : (Perfect gas) ; A gas which obeys all gas laws perfectly under all conditions. Practically no gas is ideal. Gases may approach ideal behaviour at high temperatures and under low pressure.

IDEAL SOLUTION : A solution of which all the components obey Rault's law.

IGNITION TEMPERATURE : The minimum temperature to which the substance must be heated to catch fire or to burn in air.

IIMENITE : FeO, TeO_2 ; A block rhombic crystalline solid occuring in sand.

IMIDES : Organic compounds possessing the group - CONHCO - e.g., phthalimide.

IMIDO GROUP : CONHCO - group.

IMINES : Compounds containing the group -NH-, e.g., dimethylamine, $(CH_3)_2$ NH.

IMINO GROUP : The divalent –NH– group.

IMPLOSION : An inward collapse of a vessel due to evacuation.

IMPROPER ROTATION : An element of symmetry involving the rotation about an axis through $2\pi n$ followed by reflection through a plane perpendicular to the axis of rotation.

INDENE : C_9H_8; A colourless, inflammable, aromatic hydrocarbon.

It has a five membered carbon ring fused to a benzene ring. A constitutent of coal tar, it is used as a solvent.

INDETERMINACY : *See Hund's uncertainty principle.*

INDICATOR : A substance used to indicate the presence of a reagent by its colour. In acid-base titrations, indicators like phenolphthalein and methyl orange are used which indicate the completion of neutralisation or end point. In fact, an indicator shows a colour change at the end point due to the very little excess of one of the reactants. Structurally, indicators are aromatic compounds. Two theories, namely Ostwald's theory and Quinoid theory are put forward to explain the functioning of indicator.

Among other indicators, starch is an important indicator for iodometry as well as iodimetry as it can form a deep blue coloured complex.

INDIGO : $C_{16}H_{10}N_2O_2$; A blue, naturally occuring dye. It is present in the leaves of plants of the genus *Indigofera* in the form of glucoside. It is manufactured synthetically nowadays.

INDIGOID DYES : Vat dyes and pigments having indole or thionaphtalene structures.

INDIUM : In; A soft, silvery element of group IIIA and V period of the periodic table. Its atomic number is 49 and relative atomic mass 114.82. It occurs with zinc blende and some iron ores. It is used in some electroplates and special fusible alloys. Some of its compounds like InAs, InP and InSb are used as semiconductors. It was discovered by Reich and Richter.

INDOLE : C_8H_7N; A compound occuring in coal tar, in plants and in the human body. It has a pyrolle ring fused to a benzene ring.

INDUCTIVE EFFECT : A permanent effect of electron shifting along a chain due to the presence of a particular group. Some groups may attract the electron density towards themselves and are called to exhibit -I effect (like NO_2 CHO, COOH, halogens etc.) while some other groups may push or release electron density towards the chain and are said to show +I effect (like CH_3, $-CH(CH_3)_2$, $-CH_2-CH_3$, $-C(CH_3)_3$ etc). A large number of properties of organic compounds can be explained on the basis of inductive effect, e.g., basic nature of amines, acidic strength of carboxylic acids, reactivities of aldehydes and ketones towards nucleophilic addition, etc. Further, stabilities of carbanions and free radicals can also be explained.

INERT GASES : *See noble gases.*

INERT PAIR EFFECT : An effect of the presence of an inert pair of electrons of ns orbital in the heavier elements of groups IIIA and IVA. The pair is not easily broken to give away electrons and so these elements prefer to exhibit lower oxidation state. For instance, thallium of group IIIA shows +1 oxidation state as a stable state, while lead of group IV shows +2 oxidation state. Their group number suggest the formation of compounds with +3 and +4 oxidation states. In the formation of compounds of higher oxidation states, an electron is to be promoted to the higher p subshell which requires more energy than compensated by the formation of extra bonds.

INFRA-RED (IR) : Electromagnetic radiations with longer wave lengths than visible radiations (more than 7800A). The rays have higher penetrating power and are invisible, used in long distance photography, I.R. lamps for relieving pain, the study of structures of organic compounds, etc.

INFRA-RED SPECTROSCOPY : IR radiation is produced by vibrational or rotational motion of molecules. These motion show characteristic absorption bands based on the presence of specific groups and the bonds present. I.R. spectroscopy provides valuable information about molecular structure, groups present and molecular symmetry. I.R. spectra, called finger prints, help in the identification of unknown compounds.

INHIBITION : A decrease in the rate of a catalysed reaction by certain other additives is called inhibition. Inhibitor generally forms a complex and thus inhibits the reaction from proceeding further. In biochemical reactions proceeding by enzymes, an inhibitor may combine with the active site of enzyme and thus prevents its normal functioning. Such substances have a toxic effect as they create hindrance in the normal body metabolism. In case of heterogeneous catalysis, compounds like H_2S, H or and some mercury salts act as inhibitors.

INK : Colouring matter which when dissolved/dispersed in a solvent may bind itself to the surface. Used in writing, printing, etc.

INNER TRANSITION SERIES : *See actinides and lanthanides.*

INORGANIC CHEMISTRY : The branch of chemistry that deals wth the study of elements and their compounds (excluding carbon compounds). Some of the simple carbon compounds like CO, CO_2, CS_2, carbonates and cyanides are also studied in it.

INSECTICIDES : Materials used to kill insects.

INSTABILITY CONSTANT : A property to measure the dissociation of a substance (usually co-ordination complexes).

INSULATORS : Substances with very low or nearly zero electrical conductance like pure ionic solids, plastics, rubber etc.

INSTRUMENTATION : The measurement of the conditions and the control of processes within a chemical plant.

INSULIN : A hormone secreted by the pancreas that promotes the intake of glucose by the tissues and so controls its concentration in the blood stream. Deficiency of this hormone in the body causes the accumulation of glucose in the blood and its subsequent excretion in the urine (diabetes mellitus). Structurally, it is a protein.

INTERATOMIC DISTANCE : The distance between the nuclei of bonded or non bonded atoms, in a molecule or crystal. It may be measured by spectroscopy or by electron diffraction method.

INTERFACIAL ANGLE-(CONSTANCY OF) : The angle between the adjacent faces of a crystal is constant and remains independent of the size of the crystal.

INTERMEDIATE COMPOUND : A compound formed within a process which cannot be isolated as a product.

INTERMETALLIC COMPOUND : A compound consisting of two or more metals present in the definite proportion in an alloy.

INTERMOLECULAR FORCES : Weak forces occuring between molecules like hydrogen-bonds, Van der Waal's forces, etc.

INTERNAL CONVERSION : A process in which an excited atomic nucleus decays to the ground state but the energy released is not emitted as a photon and is transferred to one of the bond electrons of that atom. This excited electron is then ejected from the atom with a kinetic energy equal to the difference between nuclear transition energy and the binding energy of the electron. Resulting ion is in the excited state and emits an Anger electron or a photon.

INTERNAL ENERGY : It is the sum of various forms of kinetic energies of all the atoms and molecules of the system and the potential energies associated with their mutual interactions. The value of the absolute internal energy cannot be measured, the significant quantity is the change of internal energy. For a closed system, the change in internal energy is equal to the difference of heat absorbed and the work done by the system.

INTERSTITIAL DEFECT : *See defects.*

INTERSTITIAL COMPOUND : A compound in which ions or atoms of a non-metal occupy interstitial positions in a metal lattice. Examples include certain carbides, borides or silicides.

INTRINSIC ENERGY : *See internal energy.*

INVAR : A trade name of an alloy of iron, nickel and carbon, having 36% nickel. It has very low coefficient of expansion over a restricted temperature range. It is used in watches and other instruments, to reduce their sensitivity to any temperature variation.

INVERSION : A reaction which involves the change from one optically active configuration to the opposite configuration.

INVARIANT SYSTEM : A term used in the study of phase diagrams which represents the state in equilibrium with no degree of freedom.

INVERSION TEMPERATURE : The temperature above which a gas is warmed up by expansion is called its inversion temperature. Most of the gases are cooled by rapid expansion. Below 80°C hydrogen is cooled but above this temperature it gets warmed up by expansion.

INVERTASE : An enzyme that helps in the hydrolysis of sucrose to glucose and fructose.

INVERT SUGAR : An equimolar mixture of glucose and fructose obtained as a result of hydrolysis. Sucrose is dextrorotatory whereas invert sugar is laevorotatory.

IODIC (V) ACID (IODIC ACID) : A colourless or light pale yelow solid, soluble in water but insoluble in pure ethanol and other organic solvents. It is prepared by the oxidation of iodine by concentrated nitric acid or hydrogen peroxide or ozone. It is a strong acid as well as an oxidising agent.

IODIC (VII) ACID (PERIODIC ACID) : A white solid, hygroscopic compound, very soluble in water, ethanol and ethers. It may be prepared by the electrolytic oxidation of iodic (V) acid at a low temperature. It is a weak acid and a powerful oxidising agent.

IODIDES : *See halides.*

IODIMETRY (IODIMETRIC TITRATIONS) : A volumetric method in which iodine is directly titrated with reagents like sodium thiosulphate. The indicator used is generally freshly prepared starch solution.

IODINE : I; A dark, violet non metal of the 17th group and 5th period of the periodic table having atomic number 53 and relative atomic mass 126.9045. A halogen, it is insoluble in water and soluble in ethanol and other organic solvents. It sublimes on heating. It is required by living organisms in very small amounts. Formerly, it was extracted from seaweeds. Its solution in ethanol is used as antiseptic and called tincture of iodine. Chemically, it is less reactive than the other halogens and it is the most electropositive halogen. It was discovered by Courtois.

IODINE (V) OXIDE : I_2O_5; A white solid, soluble in water, acts as an oxidising agent. It decomposes at high temperatures. Its solution in water is iodic (V) acid.

IODINE VALUE : A measure of the amount of unsaturation in a fat or vegetable oil. It is obtained by finding the percentage of iodine by weight absorbed by the sample in a given time under standard conditions.

IODOETHANE : (ETHYL IODIDE) ; C_2H_5I; A colourless liquid prepared by reacting ethanol with a mixture of iodine and red phosphorous, or potassium iodide and phosphoric acid. It may also be obtained by treating ethene with hydrogen iodide.

IODOFORM : *See triiodomethane.*

IODOFORM TEST : *See haloform reaction.*

IODOMETHANE : (Methyl Iodide) ; CH_3I; A colourless liquid made by reacting methanol with a mixture of iodine and red phosphorous, or methanol with a mixture of potassium iodide and phosphoric acid.

ION : An atom or group of atoms having some positive or negative charge. Positively charged ion, called cation, is formed by the loss of one or more electrons, while a negatively charged ion, called anion, is formed by the gain of one or more electrons. Ions are formed by heterolytic fission of a bond also.

ION EXCHANGE : The exchange of ions by a suitable solid in contact with a solutions usually aqueous. Applications of fertilisers is based on this principle. When potassium chloride is given to the soil, it absorbs potassium ion and releases sodium and calcium ions. Synthetic ion exchangers are copolymers having a cross-linked three dimensional structure to which ionic groups are attached. A cationic resin has positive ions built in the structure and exchange negative ions. Inorganic polymers like zeolite also act as ion exchangers. In it, positive ions are held at sites in the silicate lattice. These are used for water softening in which calcium ions in the water displace sodium ions in zeolite. The zeolite can be regenerated with concentrated sodium chloride solution. Ion exchange membranes are used to remove salts from saline water and in producing deionised water.

Ionic Atmosphere : According to Debye Huckel's theory of strong electrolytes, ions carrying one type of charge remain surrounded by an atmosphere of ions carrying opposite charge. When an electric field is applied, the central ion moves in one direction while ionic atmosphere due to opposite charge moves in opposite direction resulting in an asymmetric distribution of ions about the central ion. The time consumed between the breaking of ionic atmosphere due to the electric field and creation of a new atmosphere is called relaxation time.

Ionic Bond : *See chemical bond.*

Ionic Crystal : *See crystal.*

Ionic Mobility : The velocity with which an ion would move under a potential gradient of 1 volt per centimeter in a solution is called absolute velocity or ionic mobility. Mathematically, it is obtained by dividing the ionic conductance by 96500.

Ionic Product : The product of concentration of ions present in a given solution, with each concentration term raised to the power equal to the number of ions made per molecule. For example, ionic product of water is $[H^+]$ $[OH^-]$ while that of silver chromate it is $[Ag^+]^2$ $[CrO_4^-]$ at a given temperature and makes the basis of pH values. Further, ionic product helps in predicting the precipitation of a substance from its solution. When ionic product exceeds the solubility product the substance gets precipitated.

Ionic Radius : Ions are considered rigid solid spheres in a crystalline solid.

X-ray diffraction method can be used to measure the internuclear distance in crystalline solids and this is the sum of the ionic radii of concerned ions. For example, in sodium fluoride this distance is 0.231 mm and this is considered as the sum of ionic radii of Na^+ and F^-. It may also be defined as the distance between the centre of the nucleus and the point to which the outermost electrons have their influence. So the values of ionic radius of same ion in different solids may be different, but by making certain assumptions about shielding effect, individual ions are assigned definite values for the ionic radii. In general, cations are smaller than their parent atoms while anions are larger than the corresponding atoms.

IONIC STRENGTH : A function expressing the effect of the charge of the ions in a solution equal to the sum of the molality of each type of ion multiplied by the square of the charge.

IONISATION : The process by which the ions are produced. Ions may be formed by the loss or gain of one or more electrons by an atom or molecule. It may require a definite amount of energy made available by radiations or heat or the reaction itself. The ease with which an atom is converted to its cation is measured by its ionisation energy. Anions are formed by the gain of electrons and this tendency is measured by electron affinity of the concerned element. Cations are formed from loss of electrons.

IONISATION ENERGY (IONISATION POTENTIAL) : The minimum amount of energy required to remove the most loosely bound electron from an isolated gaseous atom. It is measured in electron volts or in joules mol^{-1}. It depends on effective nuclear charge, total number of orbits, size, screening effect, penetration effect, etc. In general, its value increases along a period and decreases down a group in the periodic table.

IONISING RADIATION : A radiation of sufficient energy to cause ionisation in the medium through which is passes. The radiations may consist of high energy particles like electrons, protons, deuterons, helium nuclei or electro-magnetic radiations of high frequency like ultraviolet, X-rays, gamma rays etc.

ION PAIR : A pair of oppositely charged ions produced as a result of dissociation of an electrolyte, e.g., dissociation of caustic soda produces a pair of Na^+ and OH^- ions.

IRIDIUM : Ir; A silvery white transition metal found in group VIII and 6th period of the periodic table having atomic number 77 and relative atomic mass 192.2. It occurs with platinum and is mainly used in alloys with platinum and osmium. It shows oxidation states of +3, +4 and +6 in its complexes. It was discovered by Tenant.

IRON : Fe; A silvery, malleable and ductile transition metal of group VIII and 4th period of periodic table having atomic number 26 and relative atomic mass 55.847. It occurs widely in nature in combined state as haemetite (Fe_2O_3), magnetite (Fe_3O_4), limonite pyrites (FeS_2) and siderite ($FeCO_3$). Pure iron exists in three forms - alpha iron, stable below 906° C with a body centred cubic crystal structure, gamma iron, stable between 906° and 1403° C with a non magnetic face centred cubic structure and, delta iron, stable above 1403° C with a body centred cubic structure. Iron, as obtained by smelting from the blast furnace, is called pig or cast iron. It may further be processed to give wrought iron and steel. It is the fourth most abundant element in the earth's crust. It is required in traces by living organisms. Iron is a very reactive metal and so is easily corroded by moist air. It shows +2 and +3 oxidation states in its compounds. +6 oxidation state is shown in ferrates (FeO_4^-).

IRON (II) CHLORIDE : $FeCl_2$; A green yellow deliquescent compound which exists as a dihydrate. Anhydrous $FeCl_2$ can be prepared by passing a stream of dry hydrogen chloride over the heated iron. The hydrated form is obtained by dissolving metal in dilute hydrochloric acid and crystallising the product from the solution.

IRON (III) CHLORIDE : $FeCl_3$; A dark brown solid, existing as the hexahydrate. Anhydrous $FeCl_3$ is prepared by passing dry chlorine over iron or steel wool. The reaction proceeds with incandexene and iron (III) chloride sublimers as black scales. The compound is rapidly hydrolysed in moist air. Its aqueous solution is acidic in nature. The compound is soluble in water as well as many organic solvents like ethanol, ether, pyridine etc. It may dimerise in vapour phase to Fe_2Cl_6. It is used as a common reagent in organic reactions. It may replace $AlCl_3$ in Friedel Craft reactions. It is used as a halogen carrier in the halogenation of benzene and other organic compounds.

It is used in qualitative inorganic analysis also.

Iron (II) Oxide : FeO; A black solid obtained by heating iron (II) oxalate. The compound has the sodium chloride structure indicating its ionic nature. It dissolves readily in dilute acids.

Iron (III) Oxide : Fe_2O_3; A red, brown or black insoluble solid. Ferric oxide occurs naturally as haematite and may be prepared by heating iron (III) hydroxide. Ferric oxide is reduced when heated with carbon, carbon monoxide, hydrogen or coal gas.

Iron Pyrites : FeS_2; *See pyrites for details.*

Iron (II) Sulphate : $FeSO_4$; Exists as hepta hydrate, called green vitriol or copperas. It is obtained by the action of dilute sulphuric acid orison fillings. The anhydrous compound is hygroscopic. It decomposes on strong heating to iron (III) oxide, sulphur dioxide and sulphur trioxide. It is readily oxidised to iron (III) sulphate in air.

Iron (III) Sulphate : $Fe_2(SO_4)_3$; A yellow hygroscopic compound obtained by heating an aqueous acidic solution of iron (II) sulphate with hydrogen peroxide. On crystallising, the hydrate. $Fe_2(SO_4)_3.9H_2O$ is formed.

Irreversible Process : *See reversible processes.*

Irreversible Reaction : *See chemical reaction.*

Isentropic Process : A process which takes place without a change of entropy. A reversible adiabatic process is isentropic.

Isobar : A curve connecting paints of equal pressure.

Isocyanate : *See cyanic acid.*

Isocyanic Acid : *See cyanic acid.*

Isocyanide : *See isonitrate.*

Isocyanide Test : *See carbylamine reaction.*

Isoelectronic : Describing compounds, molecules, atoms or ions having same number of valence electrons. For example, CO and N_2 are isoelectronic molecules while F^-, O^-, Ne, Na^+ and Mg^{++} are all isoelectronic species.

Isoleucine : *See amino acid.*

Isomerase : Enzymes used to catalyse the isomerisation of compounds.

ISOMERISM : Compounds which have the same molecular formulae but different molecular structure or different arrangements of atoms and groups in space are called isomers and this phenomenon is called isomerism. It is of two types - structural and stereo isomerism. In structural isomerism, the molecules have different structures. It is further of various types, like chain, position, functional, metamerism, tautomerism etc. Structural isomers have generally different physical and chemical properties. In stereoisomerism, the isomers differ only in the arrangement of groups in space. It is of two types - optical isomerism and geometrical isomerism.

ISOMERS : *See isomerism.*

ISOMETRIC : A curve on a graph relating temperature and pressure at constant volume or denoting a system in which the axes are perpendicuar to each other.

ISOMORPHISM : Substances showing the same crystal structures so that on mixing, they are able to form a solid solution.

ISONITRILE : (Isocyanide, Carbylamine); An organic substance having the group - N =C.

ISO-OCTANE : *See octane number.*

ISOPRENE : $CH2=C(CH_3)CH=CH_2$; 2-methyl- 3- butadine. It is the monomer of natural rubber as well as a structural unit in terpenes.

ISOSTRUCTURAL : Compounds,substances or elements having the same type of lattice and crystal structure.

ISOTACTIC POLYMERS : *See polymers.*

ISOTHERMAL PROCESS : A process that takes place at constant temperature. In such a process, heat may be evolved or absorbed without any change in the temperature.

ISOTONES : Atomic nuclei having same number of neutrons but different number of protons and different mass numbers.

ISOTONIC : Describing solutions having the same osmotic pressure values, i.e., the solutions which have same concentration and same molecular state in solutions.

ISOTOPES : Two or more atoms of the same element that have the same number of protons in their nuclei but different number of neutrons. Thus, they occupy the same place in the periodic table although mass numbers of these atoms will be different. Most of the elements in nature consist of several isotopes. For example, hydrogen exists as a mixture of H^1, H^2 and H^3, carbon exists as a mixture of C^{12} and C^{13} oxygen exists as a mixture of O^{16} and O^{18} etc. The isotopes differ in physical properties because of difference in mass. This property is used for their separation. Other methods for their separation include fractional distillation, exchange reactions, diffusion etc.

Isotopic Mass : The mass number of a given isotope of an element.

Isotopic Number : The difference between the number of neutrons and protons in an isotope.

Isotropic : Describing a medium whose physical properties are independent of direction, e.g., amorphous substances.

J

Jade : A hard semiprecious stone consisting of jadiete or nephrite. Jadiete is sodium aluminium pyroxene $NaAlSi_2O_6$. Its importance lies in its intense translucent green colour. Nephrite is one of the amphiboles, it occurs in a variety of colours like green, yellow, white, black, etc.

Jadiete : *See jade.*

John-Teller Effect : If a likely structure of a non linear molecule or ion would have degerate orbitals the actual structure is distorted so as to split the energy levels. The effect is observed in inorganic complexes. For example, the ion $[Cu(H_2O)_6]^{2+}$ is octahedrral and the six ligands are expected to occupy equidistant positions, in fact, it is distorted with four ligands showing a square geometry but the remaining two ligands opposite to each other are slightly more away from the metal. In such a case distorted geometry, there is also a centre of symmetry. The effect was put forward by H.A. John and Edward Teller.

Jasper : An impure form of chalcedony. Due to iron oxide impurities, its colour is red or reddish brown. It is used as a gemstone.

Jelutong : A rubber like material obtained from the tree *Dyera costularia.*

Jet : A variety of coal which may be cut and polished and so is used in making ornaments and jewellery.

Jeweller's Rouge : Red powder used as mild abrasive to clean metals. It is actually powdered haegnetite.

JONES REDUCTOR : A tube having zinc amalgam, used to reduce solution prior to its estimation.

JOULE : J; The S.I. unit of energy of work. It is equal to the work done when the point of application of a force of one Newton moves in the direction of the force, a distance of one metre. It is related to other units as : 1J = 10^7 ergs = 0.2388 calories. Named after the scientist James Prescott Joule.

JOULE'S LAW : The internal energy of a given mass of gas is a function of its temperature and does not depend on its volume or pressure. The law applies only to the ideal gases. In case of real gas, intermolecular forces of attractions change with volume and so influence internal energy also.

JOULE-THOMSON EFFECT (JOULE-KELVIN EFFECT) : Whenever a gas expands through a porous plug into a region of low pressure the temperature falls. It is due to the fact that during expansion, some work has to be done by the gas to overcome intermolecular forces. It is in fact a state of deviation from Joule's law, according to which there should be no change in temperature with increase of volume. At a given pressure, there is a definite temperature (inversion temperature) above which the gas exhibits increase of temperature on expansion. Below this temperature, all gases show cooling. Hydrogen and helium exhibit heating on expansion as their inversion temperatures are very very low. The phenomenon is used for liquifying air as well as making solid carbon dioxide (dry ice).

K

KAINITE : $MgSO_4$ KCl. $3H_2O$; A naturally occuring double salt of magnesium and potassium used as a fertiliser.

KALINITE : $Al_2(SO_4)_3$. K_2SO_4. $24H_2O$; A mineral form of aluminium potassium sulphate.

KAOLIN : (China clay); A soft white clay which is a composed of mainly kaolinite. It is formed during the weathering of clays or feldspar. It is used as raw material in the ceramics industry, as a filler in the manufacture of rubber, paper, paint and textiles, etc.

KARL FISCHER REAGENT : A mixture of iodine and sulphur dioxide dissolved in pyridine-methanol. Used in titrations.

KATHAROMETER : An instrument used to detect the presence of a small amount of an impurity in air and as a detector in gas chromatography. It is also used to compare the thermal conductivities of two gases by comparing the rate of loss of heat from heating coils surrounded by gases.

KEKULE STRUCTURE : A proposed structure of benzene having a hexagonal ring of carbon atoms linked by alternate single and double bonds. It was first proposed by F.A. Kekule.

KELVIN : S.I. unit of temperature. Temperature on Kelvin scale is equal to the temperature on degree centigrade scale plus 273.15 (generally taken as 273). Absolute zero temperature means 0K and so freezing temperature of water on this scale is 273K while boiling temperature is 373K. The unit is named after Lord Kelvin.

KELVIN EFFECT : *See Thomas effect.*

KERATINS : A group of fibrous proteins occurring in hair, feathers, horns etc. They have coiled polypeptide chains which combine to form supercoils of several polypeptides lined by sulphate bonds between cysteine amino acids.

KERNEL : A postively charged ion formed by the removal of all the valence electrons of an atom (usually a metal). In metals, kernels are bridged together by mobile or delocalised electrons and most of the properties of metals are due to the delocalised electrons.

KERNITE : An important mineral form of boron.

KEROSINE : A mixture of hydrocarbons, obtained by the refining of petroleum. *See petroleum for details.*

KETALS : Organic compounds formed by the addition of an alcohol to ketone (similar to acetals). If one molecule of alcohol is added, ketone gives hemiketal. Reaction of one more molecule produces ketal, $RR'C(OR'')_2$

KETENE : $CH_2=C=O$; A colourless gas obtained by passing acetylene through metallic tubes at a high temperature of about 823K to 1073K. Used as an acetylating agent.

KETO-ENOL TAUTOMERISM : A form of isomerism in which two isomers may be converted into one another by the shift of a proton and π-electron density.

KETOFORM : *See keto–enol tautomerism.*

KETOHEXOSE : *See monosaccharide.*

KETONE BODY : Any of the three compounds, acetoacetic acid (CH_3COCH_2COOH), β-hydroxy butyric acid ($CH_3CH(OH)CH_2COOH$) and acetone produced by the liver as a result of metabolism of body fat deposits. They are used as energy sources by peripheral tissues.

KETONES : Organic compounds containing $RC=OR$ group linked to two alkyl groups. They are called alkanones, eg. CH_3COCH_3 propanone. They are made by the oxidation of secondary alcohols by acidic $KMnO_4$ or acidic $K_2Cr_2O_7$. They may also be prepared by the hydration of alkynes other than ethyne, by ozonolysis of branched alkenes, by the hydrolysis of vicinal dihalides, by heating calcium salts of carboxylic acids, etc. They undergo nucleophilic addition reactions like addition of

HCN, RMX, ROH, $NaHSO_3$ etc. They undergo condensation reactions to yield oximes, hydrazones, phenylhydrazones, semicarbazones, etc. They are not easily oxidised and so do not reduce Tollen's reagent or Fehling is solution. Under drastic conditions they may be oxidised to form carboxylic acids having lesser number of carbon atoms, e.g., butanone on oxidation, produces ethanoic acid.

KETOPENTOSE : *See monosaccaharide.*

KETOSE : Carbohydrates having a ketone group. *See monosaccharides.*

KETOXIMES : Organic compounds having the group =C=NOH formed by the action of hydroxylamine on a ketone. On reduction they form primary amines.

KIESELGUHR : A soft grained deposit consiting of the siliceous skeletal remains of diatoms formed in lakes and ponds. It is used as an absorbent, filler and insulator.

KIESERITE : A mineral form of magnesium sulphate monohydrate, $HgSO_4.H_2O$.

KILO : K; A prefix used in the metric system denote 10^3 e.g., 1000g = 1Kg.

KILOGRAM : Kg; The S.I. unit of mass defined as a mass equal to that of the international Pt-Ir prototype kept by the International Bureau of Weights and Measures at Sevres, Paris.

KILOWATT-HOUR : A unit of electrical energy equal to the energy transferred by one kilowatt power per hour (1 Kwh = 3.6×10^6J).

KIMBERLITE : A rare igneous rock that often contains diamonds. It consists of olivine and phologpite mica, usually with calcite, serpentine type of minerals.

KINASE : An enzyme used to catalyse the transfer of phosphate from ATP to the final substrate.

KINEMATIC VISCOSITY : D; The ratio of the viscosity of a liquid to its density.

KINETICS : A branch of chemistry which deals with the study of rates of reactions and their dependence on physical conditions like concentration, temperature, pressure etc.

KINETIC THEORY : A theory which explains the physical properties of matter in terms of the motion of its constituent particles. Kinetic theory of gases may explain almost all the properties of gases like compressibility, liquefaction, diffusion, energy etc. Various postulates of theory may be given as :

(a) The gas molecules are considered rigid, solid, spherical point particles which are always in regular random motion.

(b) Gas molecules collide with one another, as well as with the walls of the vessel.

(c) Collisions on the walls are responsibile for pressure.

(d) Actual volume of the gas molecules is negligibly small as compared to total volume occupied by the gas.

(e) Collisions are perfectly elastic, so no energy is lost during collisions.

(f) Mathematically, by applying kinetic theory to the gaseous system, an ideal gas obeys the equation, $PV = 1/3mN\mu^2$, where m is mass of the molecule N is the number of molecules, μ is the root mean square velocity. With the help of this equation various gas laws like Boyle's law, Charles' law, Graham's law, etc. can be derived.

In case of liquid, molecules are still in random motion but due to the closeness of molecules with one another, attractive forces operating among the molecules become significant.

In case of solids, the atoms, ions or molecules are so tightly held in the crystal lattice that free motion is not possible.

KIPP'S APPARATUS : An apparatus to produce a uniform supply of gas. In the laboratory it is used to prepare hydrogen sulphide gas by the action of iron sulphide and dilute sulphuric acid. It is named after Petrus Kipp.

KIRCHOFF'S EQUATION : An equation derived from the first law of thermodynamics relating the specific heats of reactants and products with the total heat.

$$\left(\frac{\delta H}{\delta T}\right)_p = \Delta Cp$$

A simple relation is obtained as under Δ :

$$\Delta H_2 - \Delta H_1 = C_p (T_2 - T_1)$$

KJELDAHL'S METHOD : A method used for the estimation of nitrogen in organic compounds. The nitrogenous compound is first converted to ammonium sulphate by refluxing with concentrated sulphuric acid in a specially made flask called Kjeldahl's flask (a long neck, round bottom flask). The mixture is then treated with caustic soda to get free ammonia which is directly passed into standard acid solution. The amount of ammonia and consequently amount of nitrogen can be determined by titrating the residual acid solution against standard alkali solution. The method was first used by the Danish chemist Johan Kjeldahl.

KNOCKING : A process responsible for the lowering of efficiency of a spark engine due to improper ratio of fuel and air or due to poor quality of the fuel used. Knocking is the undesirable metallic voice produced in the engine which damages the piston and cylinder. It is decreased by the addition of some antiknocking agents like tetraethyl lead and ethylene dibromide.

KOHLRAUSCH'S LAW : The conductivity of a diluted salt solution is the sum of two values - one depending on cations and the other on anions. The law is based on the independent migration of ions at infinite dilution. The law was experimetnally derived by the German chemist Friedrich Kohlrausch.

KOLBE'S METHOD : A method of preparing alkanes by the electrolysis of aqueous solutions of sodium or potassium salts of corresponding carboxylic acids. For example, ethane is obtained by the electrolysis of sodium acetate.
The method was discovered by the German chemist Herman Kolbe. The method can now be applied for preparing ethene (by electrolysing potassium succinate) and ethyne (by electrolysing potassium fumarate).

KOVAR : A trade name for an alloy of iron, cobalt and nickel, which is similar to glass. It is used in making glass to metal seals especially in circumstances in which the temperature may vary.

KROLL PROCESS : A method for producing certain metals by reducing the chloride with magnesium metal. For example, preparation of titanium chloride ($TiCl_4$) with magnesium metal.

KRYPTON : Kr; A colourless, gaseous monoatomic inert element of zero group and 4th period having atomic number 36 and relative atomic mass of 83.80. It occurs in air in very small amounts and can be extracted by fractional distillation of liquid air. The element is not isolated but is used with other inert gases in flourescent lamps and in photographic flash lights.

KUPFER NICKEL : A naturally occuring form of nickel arsenide. NiAs.

KURCHATONIUM : Ku; An element of d-block having atomic number 104. Properties have not been studied fully.

L

Label : A stable or radioactive nuclide used to investigate some processes.

Labelling : The process of replacing a stable atom in a compound with a radioisotope of the same element to enable the study of reaction mechanism. It may also be called radioactive tracing. It is widely used in chemistry, biology, medicine and engineering. For example, mechanism of esterification can be studied by this technique by taking labelled acetic acid as $CH_3\,CO^{18}OH$. It is found that the water produced is $H_2{}^{18}O$, i.e., the oxygen in the water; comes from the acid and not from the alcohol.

Labile : Describing the compound in which certain atoms or groups can be easily replaced. It is more important in case of coordination complexes in which ligands can be replaced by other ligands present.

Lactams : Cyclic organic compounds having -NHCO- group as part of the ring. They exist in two tautomeric forms : lactum and lactim. In the lactim form, the group is -N = C(OH)-. Uracil is an example of a lactam.

Lactate : An ester of lactic acid.

Lactic Acid : $CH_3CH(OH)COOH$; 2-hydroxypropanoic acid. A colourless, odourless, hygroscopic aliphatic acid. It is prepared by the hydrolysis of actetaldehyde cyanohydrin [CH_3 CH(OH) CN]. It is used in the dyeing and tanning industries. It is produced from pyruvic acid in the active muscle tissues.

During hard exercise, it may build up in the muscles causing cramp–like pain.

LACTIM : *See lactams.*

LACTONE : A type of organic compound containing -OCO- group as part of a ring in the molecule. It is usually formed as an internal ester when 3-hydroxy or 4-hydroxy acids are heated strongly.

LACTOSE : $C_{12}H_{22}O_{11}$; Milk sugar. A disacharide found in milk which on hydrolysis produces glucose and galactose.

LADENBURG STRUCTURE (OF BENZENE) : A structure proposed by Albert Ladenburg in which the six carbon atoms were arranged at the corners of a triangular prism, linked by single bonds to each other and to the hydrogen atoms. Structure is not valid as it does not account for its aromaticity.

LAEVOROTATORY : Describing a chemical compound which rotates the plane polarised light towards the left. It is denoted by l or - sign.

LAEVULOSE : *See fructose.*

LAKE : A pigment made by the combination of an organic dye with an inorganic compound.

LAMELLAR SOLID : A solid inorganic substance with a crystal structure made of layers. Silicates form many such compounds, e.g., talc, pyrophyllite, mica, graphite, etc. Intercallation compounds are also lamellar formed by the interposition of atoms or ions between the layers of an existing compound.

LAMP BLACK : A finely divided form of carbon obtained as a result of incomplete combustion of organic compounds. Used as a black pigment.

LANOLIN : An emulsion of a fat in water containing cholesterol and some terpenes, alcohols and esters. It is used in making cosmetics and creams.

LANSFORDITE : A mineral form of magnesium carbonate pentahydrate.

LANTHANIDE CONTRACTION : *See lanthanides.*

LANTHANIDES : Lanthanoids; Lanthanones; Rare earth elements.

A group of elements in which electrons are being filled in (n-2)f subshell. These are fourteen in number and they follow lanthanum. Their atomic numbers are 58 to 71. Lanthanum itself has no electron but is placed with these due to the similarity in properties. Their general oxidation state is +3 although some exhibit +2 and +4 oxidation numbers also. For example, Ce^{4+}, Eu^{++} and Y^{2++} are more stable forms. The f orbitals in these atoms are not very effective in shielding the outer electrons from the nucleus. In going along the series, the increasing nuclear charge causes a decrease in the radius of M^{3+} ions. It is called lanthanide contraction. This explains the similarity between the pairs of elements like zirconium and hafnium, niobium and tantalum, molybdenum and tungsten. Properties of lanthanides are quite similar and so they cannot be easily separated from one another. They may be separated by sophisticated solvent extraction and chromatographic methods.

LANTHANUM : La; A soft, ductile, malleable, transition metal of group IIIB and 6th period. It has an atomic number 57 and relative atomic mass 138.91. It is considered as the first member of the lanthanide series as it occurs with them. Main ores which contain lanthanum are monazite and bastnasite. It is used in several alloys, as a catalyst in the cracking of crude oil. Its oxide is used in optical glasses. The element was discovered by C.G. Mosander.

LAPIS LAZULI : A blue rock which is widely used as a semiprecious stone. It is composed of a deep blue mineral called lazurite embedded in white calcite.

LASER : Light amplification by stimulated emission of radiation. A laser device produces high intensity monochromatic beams of light.

LATENT HEAT : The heat evolved or absorbed when a substance changes its phase, from solid to liquid (latent heat of fusion) or from liquid to vapour (latent heat of vaporisation) at a constant temperature like melting point or boiling point respectively.

LATTICE : A regular three dimensional arrangement of points representing atoms, ions or molecules, i.e., the constituent particles of a crystalline solid. It can be examined by X-ray studies.

LATTICE ENERGY : The energy released when ions of opposite charges are brought together from infinity to make one mole of the concerned crystal. It is the measure of stability of crystalline solid w.r.t. its stability in the gaseous phase.

LAUGHING GAS : N_2O; *See dinitrogen oxide.*

LAURIC ACID : *See dodecanoic acid.*

LAW OF CONSERVATION OF ENERGY : *See conservation of energy, law of.*

LAW OF CONSERVATION OF MASS : *See conservation of mass, law of.*

LAW OF CONSTANT COMPOSITION : *See constant composition, law of.*

LAW OF DEFINITE PROPORTIONS : *See definite proportions, law of.*

LAW OF MASS ACTION : *See mass action, law of.*

LAW OF OCTAVES : *See Newland's law.*

LAWRENCIUM : Lr; Last element of actinide series, radioactive and transuranic, having atomic number 103. It is not found in nature but several of its shortlived isotopes have been synthesised in the laboratories.

LAWS OF CHEMICAL COMBINATION : *See chemical combination, laws of.*

LEACHING : Extracting an ore or metal from impurities with the help of a suitable solvent. For example, bauxite ore (of aluminium) is concentrated by washing it well with caustic soda solution.

LEAD : Pb; A grey, soft, ductile metal of group IV A and 6th period. It has atomic number 82 and relative atomic mass 207.19. Its important ore from which it is extracted is galena (PbS). Other ores include anglesite ($PbSO_4$), litharge (PbO) and cerrusite ($PbCO_3$). Lead is extracted by roasting the well concentrated ore to get oxide which is then reduced by coke.

Lead has a variety of uses like in lead accumulators, building construction, bullets, shots, solder, bearing metal, type metal, petrochemical industry, paint, glass etc.

LEAD (II) ACETATE : *See lead (II) ethanoate.*

LEAD ACCUMULATOR : A secondary cell used to convert chemical energy to electrical energy. It is used as a vehicle battery. In the electrolyte, dilute sulphuric acid (about 1.30g cm^{-3}), lead plates are dipped which act as anode. On the other hand, PbO_2 placed in lead alloy grid acts as cathode. When fully charged, its emf is about 2 volt. Usually six such cells are connected in the battery. When the reactions continue, concentration of the electrolyte falls, and the cell requires charging when the concentration of electrolyte falls to 1.20 g cm^{-3}. Reactions of discharge (i.e. functioning of the cell) are as under :

$Pb + SO_4^{-} \rightarrow PbSO_4 + 2e^{-}$ (anode reaction)

$PbO_2 + 4H^{+} + SO_4^{-} + 2e^{-} \rightarrow PbSO_4 + 2\,H_2O$ (cathode reaction).

During charging the reactions on two electrodes are reversed as the current is being passed through the cell in an opposite direction.

LEAD (II) CARBONATE : $PbCO_3$; A white poisonous powder compound which occurs in nature as the mineral cerrusite. May be prepared in the laboratory by the addition of ammonium carbonate solution to lead (II) nitrate or lead (II) acetate.

LEAD (II) CARBONATE HYDROXIDE : $2\,PbCO_3.\,Pb(OH)_2$; White lead. Basic lead carbonate. A powdery substance insoluble in water. It is prepared by the electrolysis of mixed solutions of ammonium nitrate, nitric acid, sulphuric acid and acetic acid, using lead electrodes. Formerly, it was used in paints but it easily gets tarnished due to the hydrogen sulphide gas present in atmosphere. Poisonous nature of the compound is also a drawback in its use.

LEAD (IV) OXIDE : Pb_3O_4; commonly known as red lead. An amorphous solid prepared by heating lead (II) oxide to 673 K. It is used in glass and p aint industries.

LEAD CHAMBER PROCESS : A process in the manufacture of sulphuric acid. Sulphur dioxide obtained by burning sulphur or iron pyrite is oxidised by air in large lead chambers using nitrogen (II) oxide (nitric oxide) as homogenous catalyst. Sulphur trioxide thus formed is dissolved in water. The method is expensive and has now been replaced by the contact process.

LEAD DIOXIDE : *See lead (IV) oxide.*

LEAD (II) ETHANOATE : $Pb(CH_3COO)_2$; A white crystalline solid, soluble in water. It exists as the anhydrous salt as well as a trihydrate and decahydrate. Trihydrate on heating to about 350 K changes into its anhydrous form but at 373 K, it loses ethanoic acid and water and changes into basic lead (II) acetate. It is used in laboratory due to its soluble nature and finds use in making several complexes.

LEAD (IV) HYDRIDE : *See plumbane.*

LEAD MONOXIDE : *See lead (II) oxide.*

LEAD (II) OXIDE : PbO; A yellow crystalline solid obtained by heating molten lead in air. It exists in two crystalline forms - litharge (tetrahedral) and massicot (rhombic). Litharge is formed when lead (II) oxide is heated above its melting point. Litharge is used in the rubber industry, in paint and varnish industry as well as used for the glazing of pottery.

LEAD (IV) OXIDE : PbO_2; A dark brown or black coloured crystalline solid prepared by the action of lead (II) oxide and potassium chlorate or potassium nitrate. It is used in the manufacture of safety matches.

LEAD (II) SULPHATE : $PbSO_4$; A white crystalline solid which occurs as the mineral, anglesite. It is almost completely insoluble in water and so may be contained by precipitating it from any lead (II) salt (like ethanoate or nitrate) by an aqueous solution containing sulphate ions. With lead (II) hydroxide and water, it forms basic lead (II) sulphate which is a useful white pigment.

LEAD (II) SULPHIDE : PbS; A shining black coloured solid which occurs in nature as galena. It is prepared by passing hydrogen sulphide gas through aqueous solution of lead (II) salt. It is used as an electrical rectifer.

LEAD (IV) TETRAETHYL : $Pb(C_2H_5)_4$; Tetraethyl lead; A poisonous, colourless liquid, insoluble in water but dissolves well in benzene, ethanol, ether and petrol. It is prepared by the action of a lead sodium alloy with ethyl chloride. It is used as an additive in petrol to increase its octane number to prevent knocking and increase efficiency. A small amount of 1,2 - dibromoethane should be mixed with it in small amounts, which may react with the lead formed (causing damage to piston and cylinder of the engine) and convert it to lead (II) bromide, a volatile compound leaving the engine with exhaust gases.

LEBLANE PROCESS : An old obsolete method for the manufacture of sodium carbonate. In this process, sodium chloride is converted to sodium sulphate by heating it with sulphuric acid and then roasting it in a rotary furnace where it is first converted to sulphide using carbon. It is now immediately converted to sodium carbonate by the action of limestone. The process was invented and developed by the French chemist, Nicolas Leblane.

LECHATELIERITE : A mineral form of silica, SiO_2.

LE CHATELIER PRINCIPLE : If a system at equilibrium is subjected to some change in conditions, it shifts in a direction so that the effect of the change applied is reduced or undone. The principle is very useful in deciding the most favourable conditions for several industrial processes. For example, the most suitable conditions for the manufacture of ammonia by Haber's process should be high concentration of reactants, nitrogen and hydrogen, high pressure and low temperature.

LECLANCHE'S CELL : A primary voltaic cell having carbon anode and zinc cathode with a 10-20% aqueous solution of ammonium chloride as electrolyte. Manganese (IV) oxide mixed with powdered coke in a porous pot surrounding the anode acts as a depolariser. Its dry form is used in torches, transistors, radios, etc.

LEWIS ACID : A substance which can accept a pair of electrons to form a coordinate bond. It is either an electron deficient atom or a molecule or a cation, e.g. BCl_3, A_1Cl_3, $FeCl_3$, H^+, Cu^{++}, Fe^{++}, etc.

LIEDIG CONDENSER : A simple laboratory condenser made of a glass tube with the surrounding jacket for the flow of cold water.

LIGAND : An ion or molecule capable of forming coordinate or dative bonds with the metal atoms or ions, resulting in coordination complexes. They may be of several types depending on the ability to form dative bonds. Monodentate ligands form only one coordinate bond while bidentate ligands form two such bonds. Examples of monodentate ligands are : NH_3, H_2O, -CN, Cl-, Br-, -OH, RNH_2, etc., while those of bidentate ligands are OOC-COO-, :NH_2CH_2 COO-, NH_2 - CH_2 - CH_2 - NH_2 etc. Polydentate ligands may form more than two dative bonds, e.g. EDTA (ethylene diamine tetra acetate) is a hexadentate ligand. Bidentate and polydentate ligands can form closed chain or ring complexes called chelates.

LIGAND FIELD THEORY : A modified form of crystal field theory in which overlapping of orbitals is taken into account.

LIGHT : A form of electromagnetic radiation visible to the eyes. Its wavelength range is between 400 mm to 700 mm.

LIGNIN : An organic polymer which is deposited within the cellulose of plant cell walls. It makes the walls rigid and thick.

LIGNITE : *See coal.*

LIME : *See calcium oxide.*

LIMESTONE : A natural form of calcium carbonate. It is used as flux and for the manufacture of cement.

LIME WATER : A solution of calcium hydroxide in water. Used as a laboratory reagent to detect carbon dioxide and carbonate.

LIMONITE : A group of hydrate iron oxides. Geothite and haemetite are important constituents alongwith silica, clay and maganese oxide. It is used as an ore of iron and as a pigment.

LINEAR MOLECULE : A molecule in which the atoms are in a straight line, e.g., O=C=O.

LINE SPECTRUM : A spectrum composed of a number of discreet lines corresponding to single wavelengths of absorbed or emitted radiations. Line spectra are the finger prints of elements and the elements can be identified with their help.

LINNZ–DONNENITZ PROCESS : A method of making high quality steel. Molten pig iron and scrap are filled into a furnace (like a Bessemer furnace). Oxygen under high pressure is blown onto the surface of the metal through a water cooled pipe. The excess heat produced enables upto 30% of scrap to be incorporated into the charge. It is also termed as basic-oxygen process.

LINOLEIC ACID : $CH_3(CH_2)_4CH=CHCH_2CH=CH(CH_2)_7COOH$; An unsaturated fatty acid abundantly present in several plant fats and oils like linseed oil, groundnut oil, soyabean oil, etc.

LINOLENIC ACID : $CH_3CH_2CH=CHCH_2CH=CHCH_2CH=CH(CH_2)_7COOH$; An unsaturated fatty acid having three double bonds. It occurs in several plant oils like linseed oil, soyabean oil etc. It is an essential oil like linoleic acid.

LINSEED OIL : A pale yellow oil which contains glycerides of fatty acids like linoleic and linolenic acids. It is used in paints, varnishes, etc., as a drying oil.

LIPASE : An enzyme secreted by the pancreas and the small intestine glands of animals. It catalyses the hydrolysis of fats into fatty acids and glycerol.

LIPID : A group of compounds occuring in living organisms which are soluble in the organic solvents like chloroform, benzene, carbon tetrachloride, etc. and insoluble in water. Lipids include oils, fats phospholipids and waxes (complex liquids) and steroids, terpenes (simple lipids which do not contain fatty acids). Lipids have a variety of functions in living organisms. They are the major components of cell walls. Waxes provide vital waterproofing for body surface. Lipids can combine with proteins to form lipoproteins (in cell membranes) with polysaccharides to form lipopolysaccharides (in bacterial cell walls).

LIPOLIC ACID : A vitamin of the vitamin B-Complex. Good sources of lipolic acid include liver.

LIPOLYSIS : The hydrolysis of strong lipids in living organisms. Long term energy reserves are generally fats and oils. When needed, lipase enzyme converts them to glycerol and the corresponding fatty acids. These are then transported to tissues and provide energy by oxidation.

LIQUATION : A process of separation of a mixture of solids (generally metals from impurities) by heating. Low melting metals can be refined by this method.

LIQUEFACTION OF GASES : The conversion of a gas into its liquid phase by decreasing the temperature (below critical temperature) and increasing the pressure.

LIQUEFIED PETROLEUM GAS : LPG; Various hydrocarbons mainly propane and butanes stored as a liquid under high pressure. It is used as a fuel. Liquefied natural gas (LNG) is a similar product but has mainly methane. Its liquefaction is difficult due to very low critical temperature (190K) but it provides a convenient form for transporting very large amounts of this gas from oil wells.

LIQUID : State of matter in which the particles are bound by weak forces of attraction. Because of weaker forces, liquids can flow and change shape. It lacks the order of the solid state. These forces are stronger in comparison to gaseous state and so volume of liquids remains fixed.

LIQUID CRYSTAL : A substance which flows like a liquid but has some order in the arrangement of molecules. Nematic crystals have long molecules aligned in the same direction but distributed randomly. Cholesteric and smectic liquid crystals have aligned molecules but in definite layers.

LITHARGE : *See lead (II) oxide.*

LITHIA : *See lithium oxide.*

LITHIA WATER : *See lithium hydrogen carbonate.*

LITHIUM : Li; A light silvery reactive metal, first member of alkali metals belonging to group IA and period II of the periodic table. Its atomic number is 3 and relative atomic mass is 6.9. Its important ores are spodumen, lepidolite, petalite, tryphilite, etc. The element is obtained by converting the ore to the chloride and then electrolysing the fused lithium chloride. It is monovalent. Due to very small size, its polarising power is maximum and so its compounds exhibit covalency. It resembles magnesium in its properties (diagonal relationship) due to similarity in size and charge to size ratio. Lithium differs from the rest of the alkali metals in several properties. It reacts with hydrogen to make LiH; with oxygen it form Li_2O; with nitrogen it gives Li_3N; with water it gives Li(OH) and with halogens it forms halides. Its carbonate and nitrate decompos on heating to produce oxide, unlike the similar compounds of other alkali metals. Phenyl lithium and lithium aluminium hydride are very important compounds in organic synthesis. Like other alkali metal compounds, lithium compounds impart a purple colour to the flame of a Bunsen burner.

Lithium Aluminium Hydride : *See lithium tetrahydride aluminate (III).*

Lithium Carbonate : Li_2CO_3; A white solid obtained by adding excess of sodium carbonate solution to a solution of lithium salt. It is soluble in water in excess of carbon dioxide, forming $LiHCO_3$. On heating strongly in a stream of hydrogen, it decomposes to give lithia.

Lithium Chloride : LiCl; A white solid obtained by the action of dilute hydrochloric acid on lithium oxide, hydroxide or carbonate. At a very low temperature, it exists as a dehydrate. It is used as a flux in welding aluminium. It is soluble in alcohols, ketones and esters. It is perhaps the most deliquescent compound.

Lithium Hydride : LiH; A white crystalline solid made by the action of metal and hydrogen at a high temperature of 773 K. It is used as a reducing agent and for preparing other hydrides. It is the most stable hydride of alkali metals. It exists as its monohydrate.

LITHIUM HYDROGEN CARBONATE : $LiHCO_3$; A compound that can exist in solution only. It is formed by the action of carbon dioxide on aqueous Li_2CO_3. The solution is used in medicine by the name lithia water.

LITHIUM HYDROXIDE : Li(OH); A white crystalline solid made by the action of metal and hydrogen at a temperature of 773 K. It is used as a reducing agent and for preparing other hydrides. It is the most stable hydride of alkali metals.

LITHIUM OXIDE : Li_2O; Lithia; A white solid made by burning lithium in air or by the thermal decomposition of its carbonate or hydroxide. It reacts slowly with water forming lithium hydroxide, releasing heat.

LITHIUM SULPHATE : Li_2SO_4; A white solid prepared by dissolving lithim oxide, hydroxide or carbonate in dilute sulphuric acid and crystalling it as monohydrate. It does not form alums with aluminium sulphate while other alkali metals form alums.

LITHIUM TETRAHYDRIDE (III) : $LiAlH_4$; Lithium aluminium hydride; A white solid made by the action of lithium hydride with aluminium chloride. It is a powerful reducing agent. It reduces aldehydes, ketones and carboxylic acids to their correspondings alcohols.

LITMUS : A natural pigment that changes colour when in contact with acids and alkalies. In alkaline medium (above pH 8.3), it is blue while in acidic medium (below pH 4.5) it is red. Though this does not give a precise method to determine pH, it gives only a broad range.

LITRE : L; a unit of volume defined as 1000 cm^3 or 1000 ml or 10^{-3} m^3. Formerly it was defined as the volume of one kilogram of water at 4°C and standard pressure. Accurately 1 litre = 1000.028 cm^3.

LOCALISED BOND : A bond in which the electrons contributing to the bond remain between the two atoms concerned. Most bonds are localised.

LOGARITHMIC SCALE : A scale of measurement in which an

increase or decrease of one unit represents a ten fold increase or decrease. Decibles and pH values are its common examples. pH, for example is the negative logarithm of hydronium ion.

LONE PAIR : A pair of electrons in the valence shell of an atom, having opposite spins. Such a pair is not shared between two atoms. Lone pair occupies similar position in space to bond pairs and so accounts for the geometry of molecules. A molecule having a lone pair acts as a Lewis base and may donate this pair to another electron deficient species.

LONG PERIOD : *See period.*

LOWERING OF VAPOUR PRESSURE : By the addition of a solute to a solvent, vapour pressure is lowered. According to Raults's law, relative lowering of vapour pressure is equal to the mole fraction of non volatile solute $p/p^o = x_2$ where p is the decrease in vapour pressure, p^o is the vapour pressure of pure solvent at that temperature and x_2 is the mole fraction of solute given by $x_2 = n_2/n_1 + n_2$. Relative lowering of vapour pressure is a colligative property which helps in determining the molecular mass of an unknown solute particularly polymers.

LOWRY BRONSTED THEORY : *See acid*

LUMEN : lm; The S.I. Unit of luminous flux, equal to the flux emitted by a point source of one candela in a solid angle of one steradian.

LUMINESCENCE : The emission of radiation from a substance when it has absorbed energy and been excited to a higher energy state. When it returns to ground state or lower energy state, energy in the form of electromagnetic radiations is emitted.

If the luminescence persists after the removal of source also, it is called phosphorescence and if luminesence disappears with disconnecting the energy source, it is called fluorescence.

LUTETIUM : Lu; Last element of lanthanide series. It is a silvery metal which occurs with other lanthanides. Its atomic number is 71 and its relative atomic mass is 174.97.

LUX : S.I. unit of illumination. It is equal to the illumination produced by a luminous flux of one lumen falling on the surface of one square metre.

LYMAN SERIES : A series of lines in the ultraviolet region emitted by hydrogen. These lines are obtained when the excited electrons fall back to the first orbit. The wavelength of radiations in the Lyman series is minimum and it can be obtained by $- R(1/1^2 - 1/n^2)$

Where R is the Rydberg's constant.

LYOPHILIC : Describing the affinity towards solvent. When the solvent is water, the word hydrophilic may be used. The term may be used for colloidal solutions like those of starch, albumen, etc. It may be used for ions or groups of molecules also. For example, -COO end of soaps is lyophilic while other end (hydrocarbon end) is lipophilic. In case of colloids, lyophilic sols are more stable and are not easily coagulated.

LYOPHOBIC : Describing repulsion for solvent. In case of water as solvent, the term hydrophobic may be used. Colloidal solutions of metals and metal salts are generally lyophobic. They are much less stable and are easily coagulated by the addition of electrolytes. Ions or groups of a molecule may also represent this character. For example, the hydrocarbon end of the soap molecule is lyophobic.

LYSINE. *See amino acids.*

M

MACROMOLECULAR CRYSTAL : A crystal made up of numerous atoms joined together by covalent bonds giving rise to a three dimensional giant structure, e.g., diamond.

MACROMOLECULE : A large molecule, e.g., natural polymer like rubber, starch, cellulose etc., or synthetic polymers like nylon, etc.

MAGNADUR : Trade name for a ceramic material used for making permanent magnets. It contains sintered iron oxide and barium oxide.

MAGNALIUM : An alloy of magnesium and aluminium. Some of these alloys contain a little of copper, tin, lead or nickel. They are in generally strong and light.

MAGNESIA : *See magnesium oxide.*

MAGNESITE : A naturally occuring mineral form of magnesium carbonate.

MAGNESIUM : Mg; A silvery metal of the second group and third period of the periodical table, having atomic number 12 and relative atomic mass 24.321. It is a reactive metal and occurs in combined state in nature. Its important minerals are magnesite ($MgCO_3$), dolomite ($MgCO_3$ $CaCO_3$) and carnallite ($MgC_{l2,}$ $KCL.6H_2O$). It is present in sea water as its chloride. The element is extracted by the electrolysis of its fused chloride. It is used in making light alloys. The metal is quite reactive. In air it forms an impervious protective oxide layer. It burns in air with an intense bright white flame. Magnesium also reacts with nitrogen, sulphur, halogens, acid etc. Organomagnesium compounds, RMgX, also called Grignard's reagents are very good synthetic tools for making a variety of organic compounds. Magnesium was first isolated by Bussy

MAGNESIUM BICARBONATE : *See magnesium hydrogen carbonate.*

MAGNESIUM CARBONATE : $MgCO_3$; A white solid occurring in nature as magnesite. It is sparingly soluble in water. It is used as a mild antacid because of its capability of reacting with dilute acids, making salts. On heating, it decomposes to its oxide. It is used as a refractory substance and mixed with asbestos it is used to cover hot pipes to minimise heat loss.

MAGNESIUM CHLORIDE : $MgCl_2$; A white solid which usually exists as its hexahydrate. On heating the hexahydrate forms magnesium oxide and hydrogen chloride gas due to the hydrolysis of chloride. Anhydrous salt may be made by evaporating an aqueous solution in an atomsphere of hydrogen chloride gas or by heating the metal in chlorine. The substance is deliquescent in nature. It is used for the production of magnesium. The wood is used in making magnesia cement, sorel cement (MgO, $MgCl_2$) in artifical leather, as a laxative.

MAGNESIUM HYDROGEN CARBONATE : $Mg(HCO_3)_2$; An unstable compound of magnesium in solid state, it exists only in solutions. It is produced by passing carbon dioxide gas through an aqueous suspension of magnesium carbonate. It causes temporary hardness in water and on heating, it easily decomposes back to carbonate.

MAGNESIUM HYDROXIDE : $Mg(OH)_2$; A white solid which is slightly soluble in water to give an alkaline solution. It may be obtained by adding caustic alkali to magnesium chloride or sulphate. It is used as an antacid.

MAGNESIUM NITRIDE : Mg_3N_2 ; Prepared by heating magnesium in nitrogen at a high temperature of about 573K. On reacting with water it liberates ammonia leaving behind magnesium hydroxide.

MAGNESIUM OXIDE : MgO; Magnesia; A white solid occurring in nature as periclase. May be prepared by heating the metal in oxygen or by the thermal decomposition of magnesium carbonate, magnesium nitrate, magnesium carbonate, magnesium nitrate or magnesium hydroxide. It is weakly basic in nature. It is used as a refractory lining for metals and glass and cement furnaces. It is used for making reflective coating on optical instruments and aircraft windscreens.

MAGNESIUM PEROXIDE : MgO_2; A white solid made by reacting sodium or barium peroxide with a concentrated solution of magnesium salt. On heating, it decomposes to release oxygen. It is used for bleaching cotton and silk.

MAGNESIUM SULPHATE : $MgSO_4$; A white solid that occurs in several hydrated forms in nature, the most common being epsom salt, $MgSO_4.7H_2O$. Another important form is kieserite, $MgSO_4.H_20$. Epsom salt may be prepared on a large scale by treating magnesium carbonate with dilute sulphuric acid. On heating to about 423K, it loses six water molecules while at about 473K, it changes into its anhydrous state. It is used in sizing, dyeing and fire proofing of cotton and silk, in leather industry for tanning, in the manufacture of fertilisers and explosives. It is used as a purgative in medicine and as an antidote for barium poisoning.

MAGNETIC QUANTUM NUMBER : m; It determines the orientation of an orbital in a magnetic field. It helps in explaining Zeeman effect of the splitting of spectral lines when the radiations are allowed to pass through a strong magnetic field. Its values depend on the value of angular momentum quantum number (l) which range from +1 to −1 through zero.

MAGNETIC SEPARATION : A method of separating an ore from impurities making use of the magnetic properties of the ore, e.g., magnetite (an ore of iron) is concentrated by this method.

MAGNETISM : The study of the nature and cause of the magnetic field arising in a substance and their interactions with other substances. Substances are divided into various classes. On the basis of their interaction with an external magnetic field diamagnetic substances experience weak repulsion by the magnetic field and it is due to the orbital motion of electrons. Such substances have no unpaired electrons. Paramagnetism is the property of substances having unpaired electrons and it arises due to the electron spin. These substances are attracted by the magnetic field. Ferromagnetism is the strongest among all and it is due to electron spins that are aligned in a definite way. Such substances are strongly attracted by the magnetic field. Magnetic measurements help in the study of bonding and geometry of complexes.

MAGNETITE : Fe_3O_4; An ore of iron. It is used as flux, as a pigment for glasses and as a lining material for furnaces. It is strongly magnetic in nature and some of its varieties called lodestone are natural magnets.

MAGNETOCHEMISTRY : Branch of chemistry dealing with the measurement and investigation of magnetic properities of compounds, particularly transition metal complexes.

MAGNETON : A unit for measuring magnetic moments. Bohr has the value of classical magnetic moment of an electron, 9.274×10^{-24} Am^2.

MALEIC ACID : Cis form of butenedioic acid.

MALIC ACID : $HOOCCH(OH)CH_2COOH$; 2 hydroxy butanedioic acid. Occurs in living organisms as an intermediate in the Kreb's cycle and also in photosynthesis. It is found especially in the juice of unripe fruit.

MALLEABILITY : A property in which a metal can be extended in all possible directions by hammering.

MALONIC ACID : $CH_2(COOH)_2$; Propanedioic acid; Prepared by the hydrolysis of cyanoethanoic acid or its ester. It decomposes on heating to form ethanoic acid. Used in organic synthesis.

MALONIC ESTER : $CH_2(COOC_2H_5)_2$; Diethyl propanedioate or diethyl malonate is a colourless liquid with a peculiar smell. Its methylene group is called active methylene as it is present between two negative groups (carboxyl groups). The compound is widely used for synthesis of various organic compounds like fatty acids, polybasic acids, α, β unsaturated acids, ketones, cyclic compound,etc.

MALONITRILE : $CH_2(CH)_2$; Used in the synthesis of vitamin B.

MALT : Cereal grains (e.g. barley) is allowed to germinate and then heated to check it. Such grains in the powdered state are called malt. It used in producing alcohol from starch.

MALTASE : An enzyme present in the digestive system of animals and in yeast which is capable of hydrolysing maltose into glucose.

MALTOSE : $C_{12}H_{22}O_{11}$; A disaccharide composed of two glucose units. On hydrolysis, it gives glucose. It is produced by the degradation of starch and glycogen. It is used in soft drinks and edibles.

MANDELIC ACID : $C_6H_5(CH(OH)COOH$. 2-hydroxy 2–phenylethanoic acid. It is prepared by the hydrolysis of benzaldehyde cyanohydrin. It is used for urinary infection.

MAGNANATE (VI) : MnO_4^{2-} ; A salt containing MnO_4^{2-} ions, e.g. potassium manganate.

MANGANESE : Mn; A grey coloured, brittle transition metal of 3d series, atomic number 25 and relative atomic mass 54.94. It belongs to group VIIB and 4th period of the periodic table. It's ores are pyrolusite (MnO_2) and rhodochrosite ($MnCO_3$). The metal is obtained by the reduction of its oxide using magnesium (Kroll process) or aluminium (Goldschmidt process). Metal reacts with cold water as well as acids to liberate hydrogen. On heating, it reacts with nitrogen. Metal shows variable oxidation state showing oxidation numbers of +2 to +7. +2 is the most stable state. The compounds in higher oxidation states act as oxidising agents. It was discovered by Scheele.

MANGANESE DIOXIDE : *See manganese (IV) oxide.*

MANGANESE (II) OXIDE : MnO; Also called mangnous oxide. It is a green solid prepared by heating manganese (II) carbonate or oxalate in the absence of air. It may also be obtained by heating the higher oxides in a stream of hydrogen. It is basic in nature. In air it rapidly oxidises to manganese (IV) oxide.

MANGANESE (IV) OXIDE : MnO_2; Called manganese dioxide It is a black solid prepared by heating manganese (II) nitrate. It occurs in the ore pyrolusite in its hydrated form. It is a strong oxidising agent. It is used as a catalyst for the preparation of oxygen in laboratory, as a depolariser in electric dry cells and in the glass industry. On strong heating, it changes into manganese(III) oxide and trimanganese tetroxide.

MANGANESE SERQUIOXIDE : Mn_2O_3; Also called manganic oxide. It is a black powder obtained by igniting manganese (IV) oxide in air at 1073 K. With concentrated alkalies, it undergoes disproportionation to form manganese (II) and manganese (IV) ions.

MANGANOUS OXIDE : *See manganese (II) oxide.*

MANNICH REACTION : Reaction of 37% formaldehyde and an amine, like dimethylamine, with an organic compound containing active hydrogen atom. It may be termed as aminoethylation of organic compounds.

MANNITOL : $C_6H_{14}O_6$; An alcohol obtained by the reduction of mannose and found in fungi and plants.

MANNOSE : $C_6H_{12}0_6$; A monosaccharide, aldohexose, isomeric with glucose.

MANOMETER : A device for measuring pressure. It is made of a simple U-shaped glass tube containing mercury. May be of open end type closed end type.

MARBLE : A mineral form of calcium carbonate formed from limestone.

MARKOVNIKOV'S RULE : A rule which governs the addition of an electrophile to an unsymmetrical alkene and predicts the products formed. It states that whenever an unsymmetrical addendum adds to the unsymmetrical alkene, the negative part adds to that carbon which has lesser number of hydrogen atoms. The rule can help in predicting the formation of 2-chloropropane from propene. It is easily explained on the basis of ionic mechanism which must form a more stable intermediate carbocation (a secondary or tertiary). Such an intermediate then takes up a negative ion or group to form the final product.

MARSH GAS : Methane produced in marshy places by the decomposition of plants.

MASS : m; A measure of quantity of the matter in a given object. S.I. unit of mass is kilogram.

MASS ACTION, LAW OF : At a given temperature, the rate of a chemical reaction is directly proportional to the product of active masses of a reactant raised to the powers equal to the number of participating molecules of each species.

The active mass is generally taken as the concentration in mol L^{-1}, but it is valid only when there are no interactions among the reacting molecules. Generally, in dilute solutions, active masses and concentration do not differ but in concentrated solutions, interactions are exhibited and so molar concentration has to be multiplied by an activity coefficient to get the active mass. The law was given by C.M. Guldberg and P. Waage. For a reaction $aA + bB \rightarrow$ products, the law gives rate $= k [A]^a [B]^b$ where k is specific reaction rate or velocity constant.

MASS DEFECT : *See packing fraction.*

MASS ENERGY EQUATION : $E = mc^2$, where E is the total energy, m is the mass and c is the velocity of light. The equation is a consequence of Einstein's Theory of Relativity. Mass is a form of energy and energy has a mass. Conversion of the mass into energy is the source of power in radioactive substance.

MASS NUMBER : *See nucleon number.*

MASS SPECTROMETER : An instrument producing ions in a gas and their analysis because of different charge/mass ratio. In an earlier method Thomson used positive ions from a discharge tube which were deflected by electric and magnetic fields at right angles to the beam of ions. Each type of ion formed a parabola on the photographic plate. It was named the mass spectrograph.

In the latest techniques, the gas is ionised by electrons. The stream of ions formed may be directed towards a particular point (detector) by the variable electric and magnetic fields. They are used for identifying compounds, analysing mixtures, determing relative isotopic compounds, and for measuring accurate relative atomic masses.

MASURIUM : A former name for technetium.

MATRIX : A continuous solid phase in which the particles of other solid substances are present, e.g., ore embedded in stones and earth.

MATTE : In the process of extracting copper an intermediate mixture of copper sulphide and iron sulphide is formed. It is called matte.

MAXWELL : Mx; A unit of magnetic flux used in the c.g.s. system. 1 Mx = 10^{-8}Wb.

MAXWELL–BOLTZMANN DISTRIBUTION : A law describing the distribution of velocities of gas molecules in a given gaseous system. At a given temperature, a very small fraction of molecules possesses very low or very high velocity while the largest fraction of molecules have velocity neither very low nor very high. This velocity is called most probable velocity. Its value is less than the average velocity. (1:1.128).

McLEOD GAUGE : A vacuum pressure gauge in which a large volume of gas in compressed to a small volume. It causes increase in pressure, which can support the fluid column at quite a higher level.

MEAN FREE PATH : The average distance travelled between collisions by the molecules in a gas. It is equal to 1/2 nd^2, where n is the number of molecules per unit volume and d is the diameter of gas molecules. It is inversely proprtional to the pressure of the gas.

MEAN FREE TIME : The average time that passes between the collisions of molecules.

MECHANISM : Study of various steps involved in the given reaction. It accounts for the process of breaking of bonds and the formation of new bonds.

MEGA : M; A prefix denoting 10^6.

MELAMINE : A white crystalline substance, $C_3H_6N_6$. It is a six membered ring compound having C and N atoms at alternate positions. On polymerisation with formaldehyde, it gives a thermosetting plastic called melamine resin.

MELANIN : A dark brown pigment of hair, eyes and skin. It is a polymer obtained by the enzymatic oxidation of tyrosine.

MELTING POINT : The temperature at which the solid and liquid phases of substance are at equilibirium at standard pressure. It is the criterion of purity of a solid.

MEMBRANE : A thin pliable sheet of tissue which can act as a boundary. It may be natural or synthetic. They are made to permit the flow of other molecules and micromolecular matter. Filtration, ultra-filtration, osmosis and dialysis are the examples of use of membranes.

MENDELEEV'S LAW : *See periodic law.*

MENDELAVIUM : Md; A transuranic radiactive element of the 5f series (or actinides) having atomic number 101. Its most table isotope is Md-258, with half life period of 60 days.

MENDIUS REACTION : The reduction of nitride to a primary amine with the help of sodium metal and alcohol.

MENTHOL : A white crystalline terpene, $C_{10}H_{19}OH$. Present in some essential oils, it has a minty smell and can be used as a flavouring agent.

MERCAPTANS : *See thiols.*

MERCURY : Hg; A liquid transition metal of group IIB and 6th period of the periodic table, having atomic number 80 and relative atomic mass 200.59. It occurs in nature as cinnabar. It is used in various scientific equipment.

MERCURY CELL : An electrolytic cell in which one or both electrodes are made up of mercury or amalgam of some metal, e.g., Weston cadmium cell. Flowing mercury electrodes are used for the production of caustic soda and chlorine from brine solution.

MERCURY (I) CHLORIDE : Hg_2Cl_2; Mercurous chloride or calomel. It is a white solid, insoluble in water, made by adding dilute hydrochloric acid to a soluble mercury (I) salt. Its colour changes into black by mixing ammonia or an alkali solution.

MERCURY (II) CHLORIDE : $HgCl_2$; Mercuric chloride or calomel. It is a white solid, insolube in water, made by adding dilute hydrochloric acid to a soluble mercury (I) salt. Its colour changes into black by mixing ammonia or an alkali solution.

MERCURY (I) OXIDE : Hg_2O; Mercurous oxide. Its existence is doubtful. If NaOH solution is added to mercury (I) nitrate, a black precipitate is formed. It is considered to be Hg_2O, but X-ray analysis has confirmed it as a mixture of HgO and Hg.

MERCURY (II) OXIDE : HgO; Mercuric oxide. A yellow solid made by the action of sodium by hydroxide and mercury (II) nitrate solution. If it is made by heating mercury in air the colour becomes red. On strong heating it decomposes to mercury.

MERCURY (II) SULPHIDE : HgS; Mercuric sulphide, vermilion. A compound occuring in nature as cinnabar and metacinnabar. In the laboratory, it may be prepared by passing hydrogen sulphide gas into an aqueous solution of mercury (II) salt. It is soluble in aqua regia only. Used as a pigment vermilion.

MESOFORM : *See optical activity.*

MESOMERISM : Another name for resonance.

.**META** : (a) Describing alternate positions, i.e., 1,3 positions on the benzene ring. (b) Certain acids made from anhydride and water, e.g., meta silicic acid, H_2SiO_3 $(H_2O + SiO_2)$.

METABOLISM : Chemical reactions taking place in the cells of living organisms.

METABOLITE : A reactant or product in metabolism.

METAL CARBONYL : A complex compound made by metal and cabonyl group, e.g., Ne $(CO)_4$.

METALDEHYDE : $C_4O_4H_4(CH_3)_4$; A solid substance obtained by polymerising acetaldehyde below 273 K. It is a tetramer, used as fuel pellets.

METALLIC BOND : It is the force with which the metal kernels (positive ions) are attached to one another through delocalised electron cloud.

METALLIC CRYSTAL : A crystal in which metal kernals have a definite lattice arrangement. Electrons which are delocalised can move through the lattice.

METALLOCENE : Metal atom coordinated to two cyclopentadieyl ions, e.g., ferrocene $(C_5H_5)_2$ Fe.

METALLOGRAPHY : The microscopic study of the structures of solids and their alloys.

METALLOID : An element having properties in between those of metals and non metals, e.g., Ge, As, Te, Si. They are semi-conductors electrically. Their oxides are amphoteric.

METALLURGY : The study of extraction of metals from their respective ores.

METALS : Elements having certain definite and distinct properties like conductance, lustre, malleability, ductility etc. Metals are generally solids present in s-block and d-block. Some of the elements of p-block are also metals. d-block elements are called transition elements while f-block elements are called rare earth metals. Metals usually have the tendency to lose electrons and form cations. They form basic oxides and hydroxides.

METAMICT STATE : An amorphous state formed by the radioactivity of crystalline substances. In fact, crystal structure disrupts due to the ejection of alpha and beta particles.

METASTABLE SPECIES : Intermediates in some reactions obtained as a result of excitation of an atom, ion or molecule.

METASTABLE STATE : A condition of a system or body in which it appears to be in stable equilibrium, but when slightly disturbed, changes into a lower energy state. Supercooled water (below 273 K) when disturbed by adding a small crystal of ice or dust, rapidly freezes.

METATHESIS : Process of double decomposition.

METHACRYLATE : A salt or ester of methacrylic acid. Polymerises to PMA.

METACRYLATE RESINS : Resins obtained by the polymerisation of methacrylic acid.

METAHCYCLIC ACID : *See 2-methylpropenoic acid.*

METHANAL : HCHO ; Formaldehyde; A colourless compound, gaseous in nature, prepared industrially by the oxidation of methanol by air using silver catalyst at 773 K. Its 40% solution is used as a preservative for biological specimens under the name of formalin. It is further used to manufacture urea - formaldehyde resins.

METHANE : CH_4; A saturated gaseous hydrocarbon, the simplest alkane which occurs in natural gas. It is an important starting material for a large number of organic compounds. In the form of natural gas it can be used as a fuel.

METHANIDE : *See carbide.*

METHANOATE : $HCOO^-$; Formate; A salt or ester of methanoic acid.

METHANOIC ACID : HCOOH; Formic acid. A liquid alphatic carboxylic acid produced by the oxidation of methanol. It is a reducing agent like aldehydes and so gives Tollen's reagent test and Fehling solution test.

METHANOL : CH_3OH; Methyl alcohol, wood spirit; A colourless liquid alcohol produced originally by the destructive distillation of wood. Now it is manufactured by the catalytic oxidation of natural gas (methane). It is used as an industrial solvent and in the plastics and pharmaceutical industry.

METHIONINE : *See aminoacids.*

METHOXY GROUP : CH_3O-group.

METHYL ACETATE : *See methyl ethanoate.*

METHYL ALCOHOL : *See methanol.*

METHYLATED SPIRIT : Ethanol is rendered unfit for drinking by adding about 9.5% methanol. It is used as a fuel and solvent.

METHYL BENZENE : $C_6H_5{-}CH_3$; Toluene; A colourless liquid aromatic hydrocarbon. It is less toxic than benzene and is more widely used as a solvent and to make TNT explosives. It is obtained from the destructive distillation of coal tar or prepared by the aromatisation of methylcyclohexane.

METHYL BROMIDE : *See bromomethane.*

METHYLENE : *See carbene.*

METHYLENE GROUP : CH_2 group.

METHYLETHANOATE : CH_3COOCH_3; A colourless liquid ester with a fragrant smell used in perfumery and as a solvent.

METHYL ETHYL KETONE : *See butanone.*

METHYL GROUP : CH_3 group.

METHYL ORANGE : An acid-base indicator which is red in acidic medium below pH 3 and yellow above pH 4.4. It is particularly useful for titrating sodium carbonate against strongs acids where phenolphthalin can not be used.

METHYLPCHENOLS : $CH_3C_6H_4OH$; Cresols; A compound having CH_3 and OH group substituted on benzene ring in piace of two H-atoms. Exists in three isomeric forms - ortho, meta and para. They are obtained by the fractional distillation of coal tar.

2-METHYLPROPENOIC ACID : $CH_2=C(CH_3)COOH$; Methacrylic acid; A white, crystalline unsaturated carboxylic acid.

METHYL RED : An acid-base indicator which is red in solution below pH 4.2 and yellow above pH 6.3.

METRE : m; S.I. unit of length.

METRIC SYSTEM : A system of units based on metre and kilogram. S.I. units, c.g.s units and m.k.s units are all different types of metric systems.

METRIC TON : *See tonne.*

MHO : *See siemens.*

MICA : Minerals having SiO_4 tetrahedral units in layers with cations and hydroxyl grouping between the layers. Some mica minerals are; muscovite $K_2Al_4(Si_6Al_2O_{20})(OHF)_4$, lepidolite K_2 $(Al\ Ti)_{5\text{-}6}$ $(Si_{6\text{-}7}\ Al_2\ O_{20})$ $(OH,\ F)_4$. Flakes of mica are used as electrical insulators and as dielectric in capacitors.

MICELLE : An aggregate of ions having a lone hydrocarbon (hydrophobic) chain. For example, when soap or a detergent is dissolved in water, micelles are formed in which non polar hydrocarbon parts point towards centre and the polar hydrophilic groups point outwards.

MICRO : A prefix used to denote 10^{-6}.

MICRON : A unit of length equal to 10^{-6} metre.

MIGRATION : The movement of a group, atom ions or a double bond within a molecule or movement of ions under the influence of an electric field.

MILK OF MAGNESIA : *See magnesium hydroxide.*

MILK SUGAR : *See lactose.*

MILLI : m; A prefix denoting 10^{-3}.

MILLIMETRE OF MERCURY : mm Hg; A former unit of presssure

1 mm Hg pressure = 133.3224 Pa.

MINERAL : A naturally occuring substance having a definite chemical composition. Examples include coal, petroleum as well compounds containing particular metals.

MINERAL ACID : Inorganic acids like hydrochloric, nitric and sulphuric acids. They are very strong and completely ionised acids.

MIRABILITE : A mineral form of hydrated, sodium sulphate.

MISCH METAL : An alloy of cerium (50%), lanthanum (25%), neodymium (18%), praseodymium (5%) and some other lanthanides. It is used as a lighter flint and is added to copper alloys to make them harder, to aluminium alloys to give more strength.

MISCIBLE : Denoting combination of substances which, when mixed, give single phase giving rise to a homogenous solution.

MITSCHERTICH'S LAW : It states that the substances crystallising in identical crystalline forms have similar chemical composition, eg., chromium oxide, ferric oxide and alumina are isomorphous so their formulae should be similar as Cr_2O_3, Fe_2O_3 and Al_2O_3. The law is called law of isomorphism.

MIXTURE : It is the combination of two or more substances (elements or compounds) mixed in any ratio. It may be homogenous (solution having single phase) or heterogenous (alloys having different phases). In a mixture, chemical properties of the components remain unchanged. Further its components are not linked by any chemical bonds and so they can easily be separated from one another.

M.K.S SYSTEM : A system based on the metre (for length), kilogram (for mass) and second (for time).

MM HG : *See millimetre of mercury.*

MOLALITY : m; A mode of representating concentration, it is defined as the number of moles of solute dissolved per 1000 g of the solvent. A molal solution contain one mole solute dissolved in a kilogram of solvent.

MOLARITY : M; Another mode of representing concentration. It is defined as the number of moles of solute dissolved per litre

(dm^3) of the solution. A molar solution contains one mole of solute per litre of the solution.

Mole : mol; The S.I. unit of amount of substance. It is defined as the amount of substance that contains the same number of particles as are contained in 0.012 kg (12g) of C-12. This quantity of particles is called as Avogadro's constant, the value of which is 6.002045×10^{23}. A mole, thus, is the amount of matter which contains Avogadro's number of particles. In case of an element, it is equal to one gram mole atoms while in a compound, it is equal to one gram molecular mass and contains one mole of molecules. For gases, it is related to volume also. One mole of a gaseous substance or vapour of a compound occupies 22.4 liters of volume under standard conditions of temperature and pressure, i.e., 273 K temperature and one atmosphere pressure.

Molecular Crystal : A crystal in which the constituent particles occupying lattice points are molecules. The forces holding the molecules are quite weak and so melting points of such solids are low. For example, iodine, dry ice (solid carbon dioxide).

Molecular Formula : The simplest formula of a compound which describes the exact number of atoms of each element present in a molecule. For example, NH_3 shows that a molecule of ammonia possesses one atom of nitrogen and three atoms of hydrogen. Molecular mass can be calculated from the molecular formula.

Molecularity : It is the number of particles (atoms, molecules, or ions) which simultaneously and effectively collide to give the products of the reaction. It is always a whole number while the order of the reaction may be zero, a whole number or a fractional number. It gives no idea of the mechanism of the reaction. For a multistep reaction, molecularity of steps is significant and molecularity of the rate controlling step (the slowest step) may be of greater significance.

Molecular Orbital : *See orbital.*

Molecular Sieve : A substance through which molecules of a limited range of sizes can pass, enabling separation. Zeolities and other metal aluminium silicates can be made with pores of constant dimensions in their structure. When a mixture is passed through, some particles get trapped in these pores while remainder mixture passed out. The trapped molecules may be recovered by heating. A modified form of molecular sieve is used in gel filteration. Siever in this case is a gel made from polysaccharide.

MOLECULAR SPECTRUM : An absorption or an emission spectrum characteristic of the molecule.

MOLECULAR VOLUME : The volume occupied by a substance per unit amount of substance (i.e. per mole).

MOLECULAR WEIGHT : *See relative atomic mass.*

MOLECULE : The smallest particle of the matter which participates in a chemical reaction and is capable of independent existence. It is made of one or more atoms of the same or different elements combined together through chemical bonds. In case of covalent compounds, there are groups of atoms held together while in ionic compounds there are no single molecules but collections of oppositely charge ions.

MOLE FRACTION : X; Ratio of number of moles of a component to the total number of moles of all the components

$X_a = \frac{n_A}{N}$, where n_A is the number of moles of component A and N is the total number of moles present in the mixture.

MOLISCH TEST : A general test of carbohydrates. In a test tube, carbohydrate solution and alcoholic alpha napthol are mixed and concentrated sulphuric acid is poured from the side of the tube. A deep violet coloured ring at the junction of the liquids confirms the presence of carbohydrates.

MOLYBDENUM : Mo; A silvery, hard transition metal of group VIB and 5th period of the periodic table, having atomic number 42 and relative atomic mass 95.94. The element is extracted from its ore molybdenite, MoS_2. It is unreactive and remains unaffected by most acids. With alkalies in fused state, it forms molybdates and polymolybdates. It is used in making several alloy steels. Is was discoverred by Scheele.

MOND PROCESS : A method of refining nickel. When impure metal is heated to about 333 K in a current of carbon monoxide, nickel carbonyl is formed which is volatile and so is easily separated. It can now be decomposed when heated to about 453 K to get pure metal. The method was introduced by Ludwig Mond.

MONEL METAL : An alloy of nickel (about 60%), copper (about 35%) and a little of iron, managanese, silicon and carbon. It is used to make acid proof vessels.

MONOATOMIC : Describing a molecule which has only one atom, e.g. noble gas molecules.

MONOBASIC : Describing an acid which has only one replacable hydrogen atom. For example, acetic acid, hydrochloric acid, nitric acid are all monobasic. Equivalent weights of these acids are equal to their respective molecular masses.

MONOCLINIC : *See crystal system.*

MONOHYDRATE : A crystalline compound having one molecule of water of crystallisation per molecule of compound.

MONOMER : A simple molecule that joins with an other, forming polymers. It is in fact the basic unit of a polymer.

MONOSACCHARIDE : Simplest carbohydrate which does not split into smaller units on hydrolysis. According to the number of carbon atoms, they are termed tetroses, pentoses, hexoses etc. Further, they are classified as aldoses and ketoses on the basis of the characteristic group as an aldehyde or a ketone. For example, glucose is an aldohexose and fructose is a ketohexose. Most of their properties can be explained by their ring structures which may be based on pyran ring of furan ring. They usually exhibit optical activity, giving rise to enantiomeric forms.

MONOTROPY : The existence of a single allotrope of an element that is always stable than the others regardless of temperature. Phosphorous shows monotropy.

MONOVALENT : Having a valency of one.

MORDANT : Inorganic oxides or salts which are absorbed on the fabric during the dyeing process. The dyestuff can now form a coloured complex with the mordant.

Morphine : An opium product. Chemically an alkaloid, it is an analgesic and narcotic, used to relieve pain but is a habit forming drug.

Mosaic Gold : *See tin (IV) sulphide.*

Moseley's Law : The frequencies of the lines in the X-ray spectra of the elements are related to the number of protons in the nuclei of concerned elements. A straight line is obtained by plotting square roots of the frequencies of lines against atomic number. Mathematically, $v = a (z - b)$, where v is the frequency of the line, while a and b are the constants characteristic of the element. The law was proposed by H.G.Moseley.

Multiple Proportions, Law Of : When two elements combine to make two or more compounds, the weights of one which combine with the same weight of the other are in a simple whole number ratio. For example, carbon and oxygen make carbon monoxide and carbon dioxide in which oxygen combining with a fixed weight of carbon bears a simple ratio of 1:2 in the two oxides.

Multiplet : A spectral line formed by more than two closely spaced lines.

Mumetal : The original trade name for a ferromagnetic alloy containing about 78% nickel, 17% iron and 5% copper, which has high permeability. Modern forms of these alloys contain chromium and molybdenum. Such alloys are used in some transformer cores. Also used for checking the effect of external magnetic field.

Muntz Metal : A form of brass containing 60% copper, 39% zinc and a little lead and iron. It is used for forging nuts and bolts, etc. Named after G.F.Muntz.

Muriate : An obsolete name for a chloride. Muriate of potash is KCl.

Muscovite : A mineral form of potassium aluminosilicate, $K_2Al_4 (Si_6Al_2) O_{20} (OH, F)4$, an important member of mica group of minerals. It is silvery grey in colour, sometimes tinted with green, brown or pink. It is also called white mica or potash mica. It is a common constituent of certain granites and pegmatites, in metamorphic and sedimentary rocks. Used in the manufacture of electrical equipment and as a filler in paints and roof material.

MUSTARD GAS : $(ClCH_2CH_2)_2$ S. Dichlorodiethyl sulphide. It is highly poisonous gas made from ethene and disulphur dichloride (S_2Cl_2). It was used as war gas in World War I

MUTAROTATION : Change of optical activity with time, as a result of spontaeous reactions. For example, glucose and glucose both initially have different angles of rotation, but on keeping for some time in solution, acquire a constant intermediate value.

MYOGLOBIN : A globular protein occuring in muscle tissue as an oxygen carrier. It comprises a single polypeptide chain and a haem group which reversibly binds a molecule of oxygen.

N

NAPLAM : A gel of gasoline obtained by adding soap of naphthalenic and palmitic acids, used in bombs.

NAPHTHA : A coal tar product containing mostly xylenes and higher homologues of benzene.

NAPHTHALENE : $C_{10}H_8$; A white crystalline solid with a distinctive smell like that of moth balls. It is obtained by the fractional distillation of middle oil and heavy oil fractions of crude oil. It has fused benzene ring structure, i.e., two benzene rings are fused together. Its properties resemble other arenes and exhibits aromaticity. It is used in the manufacture of phthalic anhydride and consequently for the manufacture of several plastics, polymers and dyes.

NAPHTHOLS : $C_{10}H_7OH$; Two phenols derived from naphthalene which differ in the position of the -OH group are called naphthol or naphthalene-1-ol or naphthol-1 and naphthol or naphthalen -z- of 1 or naphthol-2. The former naphthol is a colourless crystalline solid, soluble in alcohol, ether and benzene. It is prepared by alkali fusion of naphthalene -1-sulphonate. It forms phthalic acid on oxidation. Forms a series of dyes on diazotisation.

Naphthol, on the other hand is a white crystalline solid soluble in organic solvents. It is prepared by caustic soda fusion of sodium naphthalene-2-sulphonate.

NASCENT HYDROGEN : H; A very reactive form of hydrogen. It is short lived and changes into normal molecular hydrogen. It cannot be separated. It is used to prepare hydrides of phosphorous, arsenic and antimony, which cannot be prepared by using ordinary hydrogen in molecular state.

NATRON : A mineral form of hydrated sodium carbonate, $Na_2 CO_3.H_2O$.

NATTA PROCESS : *See Ziegler process.*

NATURAL ABUNDANCE : *See abundance.*

NATURAL GAS : Gas present in association with petroleum or sometimes independently below the rocky strata of earth. It has a very high percentage of methane (85%), along with ethane (10%), propane (3%) and other hydrocarbons in very small proportions. Some other gases like CO_2, CO, H_2, etc. may also be present. In certain countries, it is directly used as a domestic and industrial fuel.

NEMATIC CRYSTAL : *See liquid crystal.*

NEODYMIUM : Nd; A silvery element of the 4f series (lanthanides) of metals existing in two allotropic forms having atomic number 60 and relative atomic mass 144.24. It occurs in bastnasite and monazite alongwith other lanthanides. It is used to colour glass violet-purple, as catalyst (in alloy form) and in carbon arc searchlights. It was discovered by C.A. Von Welsbach.

NEON : Ne; A colourless odourless gas belonging to zero group of periodic table, having atomic number 10 and relative atomic mass 20.18. It occurs in minute quantities in air. It is used in neon sign lights, lamps and other electrical equipment. The element was disvovered by Sir Ramsay and Travers.

NEOPRENE : A synthetic rubber produced by the polymerisation of chloroprene. It is more resistant to oil solvents, high temperature and chemicals.

NEPTUNIUM : Np; A silver coloured inner transition element of the actinide series having atomic number 93 and relative atomic mass 237.0482. It is highly radioactive and the first transuranic element to be synthesised. Its most stable isotope is Np-237, having a half life period of 2.2×10^6. It was discovered by McMillan and Abelson.

NEPTUNIUM SERIES : *See radioactive series.*

NERNST HEAT THEOREM : It states that there is no change of entropy when a chemical change takes place between pure crystalline solids at absolute zero.

NESQUEHONITE : A mineral form of hydrated magnesium carbonate, $MgCO_3.3H_2O$.

NESSLER'S REAGENT : K_2HgI_4; A solution obtained by adding excess of potassium iodide to an aqueous solution of mercury (II) chloride. An orange coloured precipitate of HgI_2 formed dissolves to give K_2HgI_4. It is used to test ammonia which gives a brown precipitate when passed through Nessler's solution.

NEUTRAL : Describing a solution which is neither acidic nor basic. Its pH value at 298 K is 7, like that of pure water.

NEUTRALISATION : The process in which an acid reacts with a base completely to form a salt and water. Neutralisation point may be detected with indicators.

NEUTRON : A neutral particle present in the nucleus having mass slightly greater than that of the proton. Its mass is 1.67492 x 10^{-27} Kg. A neutron may change (decay) into a proton, an electron, an antineutrino within the nucleus. It was discovered by James Chadwick.

NEUTRON NUMBER : N; The number of neutrons present in the nucleus of an atom.

NEWLAND'S LAW (OF OCTAVES) : When the elements are arranged in the order of increasing atomic weights, there is similarity between the elements that have a gap of seven. For example, Li resembles Na; Be resembles Mg. The law was discovered by Newland before the periodic law of Mendeleev came to light.

NEWTON : N; Unit of force. 1 N = 1 Kg ms^{-2}. It is defined as the force needed to accelerate one kilogram mass by one metre per second square.

NIACIN : *See nicotinic acid.*

NICHROME : A trade name for a group of nickel-chromium-iron alloys containing 60-80% nickel, about 16% chromium and small amounts of other elements like iron, carbon, silicon etc.

Due to very good resistance to heat and oxidation, they are used for making wires of heating elements.

NICKEL : Ni; A malleable ductile silvery metal of the 3d series of transition elements. Occurs as sulphides. Its atomic number is 28 and its relative atomic mass in 58.70. In nature, it occurs in the mineral penthandite, NiS. The metal is extracted by roasting the ore to give oxide followed by reduction with carbon monoxide. The metal is purified by Mond's process. It is used as a catalyst in the hydrogenation of unsaturated hydrocarbons, oils, etc. It is used in special steels, Invar, magnetic alloys etc. The main oxidation state is +2. Carbonyl complexes exhibit zero oxidation state. It was disovered by A.F. Cronstedt.

NICKEL CARBONYL : $Ni(CO)_4$; A colourless, volatile toxic liquid prepared by passing carbon monoxide gas over finely divided nickel at about 60°C. The reaction can be reversed at a high temperature, so this becomes the basis for Mond's process.

NICKELLIC OXIDE : *See nickel (III) oxide.*

NICKEL-IRON ACCUMULATOR : (Nife cell or Edison cell). It is a secondary cell devised by Thomas Edison which has an iron cathode and nickel oxide as anode. Each cell gives an e.m.f. of about 1.2 V when potassium hydroxide solution forms the electrolyte.

NICKEL (II) OXIDE : NiO; nicklous oxide. A light green powder formed by heating nickle (II) hydroxide, nitrate or carbonate in the absence of air. It may be reduced to nickel by carbon, carbon monoxide or hydrogen.

NICKEL (III) OXIDE : Ni_2O_3; Nickelic oxide. A black powder prepared by heating nickel (II) oxide in air. It exists as its dihydrate.

NICKEL-SILVER : (German silver); Alloy containing nickel, copper and zinc in different proportions. It is used in silver plating, chromium plating, and enamelling.

NICOTINE : A colourless alkaloid, toxic in nature, present in tobacco.

NICOTINIC ACID : (Niacin); A vitamin of the b-Complex group, manufactured from amino acid tryptophan by plants and animals.Its deficiency causes the disease pellagra in man. Its good sources are liver, groundnut and sunflower oils.

NIFE CELL : *See nickle-iron accumulator.*

NINHYDRIN : A colourless organic compound which gives blue colour with amino acids. It is a test for amino acids.

NIOBIUM : Nb; A soft ductile grey-coloured transition element of the 4d series, having atomic number 41 and relative atomic mass 92.91. It occurs in nature in the mineral niobite, $Fe(NbO_3)_2$. It is used in special steels and in welded joints. Nb-Zr alloys are used in superconductors. The element was called columbium in the beginning. It was discovered by Charles Hatchett.

NITRATE : NO_3; A salt or ester of nitric acid.

NITRATION : Introduction of nitro group ($- NO_2$) into an organic compound. In case of aromatic compounds, it involves the use of a mixture of concentrated nitric acid and concentrated sulphuric acid. The reaction here is electrophilic substitution and involves the use of the electrophile NO_2^+, nitronium ions.

NITRE : *See potassium nitrate.*

NITRIC ACID : HNO_3; A colourless, fuming, corrosive and strong acid prepared by the catalytic oxidation of ammonia (Ostwald's process) or by the conversion of air to nitric oxide at a very high temperature (Birkland - Eyde process). Nitric oxide thus formed is then converted to nitric acid in the presence of air and water. Most of the metals react with it forming their nitrates with the evolution of oxides of nitrogen. Reactions vary with the concentration of the acid. Some of the non metals like S, P, As, etc. react to form corresponding oxyacids. Nitric acid is used to prepare organic nitro compounds which are used in explosives.

NITRIC OXIDE : NO; *See nitrogen monoxide.*

NITRIDE, RUBBER : Synthetic rubber produced by the polymerisation of buta–1,3 diene ($CH_2=CH–CH=CH_2$) with acrylonitride ($CH_2=CH–CN$). It is more resistant to oil and solvents.

NITRIDES : Compounds having N^{3-} ions, e.g., B N, Mg_3 N_2, Li_3N. They are formed by the direct reaction between the electropostive element with nitrogen. Transition metals form several interstial nitrides like Mn_4N, W_2N etc.

NITRILES : (Cyanides); Compounds having -CN group. Usually they are colourless, pleasant smelling liquids. They are formed by treating an alkyl halide with potassium cyanide dissolved in alcohol. On hydrolysis, they form amides and corresponding carboxylic acids.

NITRITE : O-N=O; A salt of nitrous acid.

NITROBENZENE : $C_6H_5NO_2$; A yellow coloured organic liquid having a peculiar odour prepared by reflexing benzene with nitrating mixture. (conc. HNO_3 and conc. H_2 SO_4) below 333K.

NITROCELLULOSE : *See cellulose nitrate.*

NITROCOMPOUNDS : Organic compounds having nitro ($-NO_2$) group attached to an alkyl or aryl group. They are formed by the nitration of a compound using nitrating mixture. Aromatic nitrocompounds are explosive in nature. On reduction under different conditions, they produce different compounds. Ultimate reduction product is an amine.

NITROGEN : N; The first element of group VA of the periodic table, having an atomic number 7 and relative atomic mass 14. It is an electronegative nonmetal, diatomic, gaseous substance which exists in air in a free state. It is separated by the fractional distillation of liquid air. In the combined state, it occurs in various nitrates mainly potassium nitrate (salt petre) and sodium nitrate (chile salt petre).

Chemically, it is largely an inert element due to high dissociation energy (940 KJ. Mol^{-1}) and reacts with some metals like Li, Mg etc., forming nitrides. It can directly combine with hydrogen to form ammonia (Haber's process) and forms a large number of oxides like N_2O, NO, N_2O_3, NO_2, N_2O_4, N_2O_5 and NO_3. Nitrogen also forms a number of binary halides such as NF_3, NCl_3, NI_3 etc. The halides, except NF_3, are highly explosive.

NITROGEN DIOXIDE : NO_2; A dark brown coloured gas having a pungent smell, produced by heating metal nitrates like copper (II) nitrate, zinc nitrate or by heating metals like Cu, Ag, Zn, Mn etc., with concentrated nitric acid. On heating, it decomposes to nitric oxide.

NITROGEN FIXATION : A process by which nitrogen of the atmosphere is changed into its compounds. It occurs due to certain bacteria present in the roots of leguminous plants. Small amounts of nitrogen is converted to nitrogen (II) oxide in the atmosphere due to lightning or during rainy season. This second process makes the basis for the manufacture of nitric acid by Birkeland - Eyde process.

NITROGEN MONOXIDE : NO; Nitrogen monoxide. It is a colourless gas, insoluble in water. It is prepared by the action of nitric acid on copper. On a larger scale, it is prepared by the catalytic oxidation of ammonia. It is a suitable oxide of nitrogen and does not decompose even upto 1273 K. At room temperature, it is changed to nitrogen dioxide by reacting with the oxygen of the air.

NITROGLYCERINE : A highly explosive compound used in dynamite. It is prepared by treating glycerol with concentrated nitric acid and concentrated sulphuric acid. The compound is in fact a trinitrate and not trinitroglycerine.

NITRO GROUP : NO_2 or $-N=O$ group.

NITRONIUM ION : *See nitryl ion.*

NITROPHENOL : $HO.C_6H_4NO_2$; Organic compounds formed by the low temperature nitration of phenols. They are of three types based on the position of nitro group. o- and p-isomers are produced directly and are separated by steam distillation, while m-isomer is produced through m-dinitro–benzene which is converted to m-nitroaniline with the help of ammonium sulphide catalyst. It is now diazotised and hydrolysed to get m-nitrophenol.

NITROSYL ION : NO^+ Ion.

NITROUS ACID : HONO; A weak acid which exists only in solutions. It is obtained by treating a nitrite with dilute mineral acid. The compound is very important industrially for diazotisation reactions which make the basis of preparation of various azo dyes.

Nitrous Oxide : N_2O; *See dinitrogen oxide.*

Nitryl Ion : NO_2^+; nitronium ion. An electrophilic species capable of attacking benzene to cause nitration. It is produced by the action of concentrated nitric acid and concentrated sulphuric acid.

NMR : *See nuclear magnetic resonance.*

Nobelium : No; A transuranic, radioactive element of the 5f series of the periodic table. Its atomic number is 102, but properties have not been studied yet.

Noble Gases : Inert or rare gases. A group of inert, gaseous, monoatomic elements belonging to zero group of periodic table. They are helium (He), neon (Ne), argon (Ar), krypton (Kr), xenon (Xe) and radon (Rn). The valence shell of these elements is either fully filled (in He) or has stable configuration of ns^2 np^6 (in all other elements). These elements have no tendency to lose or gain or share electrons and so they do not form bonds. The only forces with which the molecules are joined are Van der Waal's forces of attraction. These elements have very high ionisation energies and zero electron affinity. Prior to 1962, no compound formation was observed in these elements. N. Bartlett was the person who prepared $Xe^+(PtF_6)$ compound and this led to the discovery of several other compounds like XeF_2, XeF_4, XeO_3, $XeOF_4$ etc. The uses of noble gases depend mainly on their inertness. Their applications are, filling light bulbs (Ar), discharge tubes (Ne), chemical research (He, Ne), sea diver's cylinders (He), therapeutics (Rn) etc. Helium is present in the atmosphere of the sun as well, as it is formed by the radioactive decay of several minerals. Argon is present upto 1% in the atmosphere. Except radon, other gases are also present in the atmosphere in minute amounts. Radon is generally obtained as a deay product of radium.

Non Benzenoid Aromatics : Aromatic compounds which have

a ring other than benzene ring, e.g., cylopentadienyl anion, C_5H_5.

NON -METALS : A group of elements which are poor conductors of electricity and heat, form covalent compounds or anions. They are on the top right hand side of the periodic table. These elements are electronegative, form acidic oxides and hydroxides. Their melting and boiling points are usually low. Some examples are N, S, P, O, F, Cl, Br, He, Ne, etc.

NON-POLAR COMPOUND : A compound which has covalent molecules with no permanent dipoles. They are either H_2Cl_2, N_2 etc. or CO_2, CCl_4 etc. In the latter category, bonds are polar but the overall resultant polarity is zero.

NON -POLAR SOLVENTS : CCl_4 Benzene, ether, acetone etc.

NON-STOICHIOMETRIC COMPOUND : A compound in which the elements do not combine in simple ratios. Such compounds are deficient in one of the constitutent elements, e.g., titanium (IV) oxide is deficient in oxygen.

NORADRENALINE : Norepinephrine; A hormone produced by the adrenal glands and some nerve endings. Most of its actions are similar to adrenaline, but it is more important to maintain normal body activity.

NORMALITY : N; The number of gram equivalents of a solute dissolved per litre of the solution.

NORMAL SOLUTION : 1 N; A solution which contains one gram equivalent of the solute per litre (dm^3) of the solution.

NTP : See STP.

NUCLEAR MAGNETIC RESONANCE : A method for identification and investigation of a compound based on nuclear spins or magnetic moments. Due to the strong magnetic field applied, splitting of nuclear energy levels takes place. Now the sample is placed in an additional, weak and oscillating magnetic field whose frequency is scanned over an appropriate range. At certain precise frequencies the nuclear magnets resonate with the field undergoing transition between the magnetic energy levels. The resonance is amplified and recorded. Hydrogen nuclei, for example, can have two energy states and the transition between the two occurs by the absorption of the radio frequency radiation. It is based on the study of various H-atoms present in different atmospheres. For example, in ethanol there are three frequencies corresponding to H-atoms of the $-CH_3$,$-CH_2$ and -OH groups. The intensity of absorption depends on the number of H-atoms of a particular kind.

NUCLEIC ACIDS : Essential constituents of all living cells, proteinous in structure, they are of two types, (a) DNA – It is made of the nucleotides of purines, adenine and guanine and the pyrimidines, cytosine and thymine. A molecule of DNA consists of two helical chains coiled round the axis, each chain comprising of alternate phosphate and sugar groups linked together to deoxyribose, a pentose molecule. They are responsible for the heriditary characteristics of an organisum.

(b) RNA – Present in the cytoplasm of a cell. The sugar present is ribose and in place of thymine of DNA, pyrimidine base is uracil. They are responsible for protein synthesis. The RNA molecules are of three types - messenger RNA, transfer RNA and ribosomal RNA.

NUCLEON : Particles present in the nucleus whose mass is 1 amu.

NUCLEON NUMBER : (Mass number); The sum of protons and neutrons present in the nucleus of an atom.

NUCLEOPHILE : An atom, group of atoms, anion or a molecule which has affinity towards the nucleus or positively changed (electron deficient) part of the molecule. Nucleophiles may be negatively charged ions or neutral molecules having one or more pairs of electrons. For example, Cl, Br, OH, CN, OR are anion nucleophiles, while NH_3, ROH, H_2O: are molecular nucleophiles.

NUCLEOPHILIC ADDITION : Addition of a nucleophile to the unsaturated part of a molecule having a double or triple bond between carbon and a highly electronegative atom like oxygen or nitrogen. It is a characteristic type of reaction in aldehydes and ketones. The addition of hydrogen cyanide, sodium bisulphate, alcohols, grignard reagents and ammonia derivatives are typical addition reactions. In some cases, a molecule of water is also eliminated from the addition product. Alkynes also undergo such addition reactions but only in the presence of heavy metal ions. For example, a water molecule adds in the presence of Hg^{++} and hydrogen cyanide adds in the presence of Ba^{++} ions.

NUCLEOPHILIC SUBSTITUTION : A reaction in which a nucleophile may replace an atom or group of atoms in an organic compound. The reaction takes place in those compounds which have a strongly electronegative atom or group of atoms. The electron deficient, slightly positively charged carbon atom is then attacked. Alkyl halides exhibit nucleophilic substitutions as very common reactions. They are of two types depending on the mechanism of the reaction (a) SN2 or bimolecular nucleophilic substitutions, in which the rate of reaction depends on both the molecular species (An alkyl halide as well as nucleophile). It is favoured in primary alkyl halide. (b) SN1 or unimolecular nucleophilic substitutions, in which the rate of the reaction depends on the concentration of alkyl halide only. In this case, a carbocation intermediate controls the kinetics. Tertiary halides undergo such type of reactions. Its reason is explained on the basis of stability of tertiary carbocation formed as intermediate, while the reaction tendency of primary halides is based on a more stable transition state.

NUCLEOSIDE : An organic compound consisting of a purine or pyrimidine base linked to a sugar, ribose or deoxyribose.

NUCLEOTIDE : An organic compound consisting of a purine or pyrimidine base linked to a sugar (pentose ribose or deoxy ribose) as well as a phosphate group. DNA and RNA may be termed as polynucleotides.

NUCLEUS : The central, heavy, dense, positively charged part of an atom. Its density is 10^{15} Kg m^{-3}. The number of protons in an atom is always definite in the given element and is called proton number. The number of neutrons may vary depending on the isotopic mass of the atom. The simplest nucleus is of hydrogen which has no neutron (in proteum), while the massive naturally occuring nucleus is of uranium. Stability of the nucleus depends on the neutron-proton ratio and binding energy. Binding energy per nucleus may be related to packing fraction as well.

NUCLIDE : A nuclear species with a given number of protons and neutrons. The word is applied to the nuclei of concerned atoms. For examples, ^{29}Cu, ^{30}Zn, ^{26}Fe are the examples of nuclides.

NUJOL : A trade name for liquid paraffin used medicinally.

NYLON : A type of synthetic polymer having peptide linkages (-NHCO-). They are generally made by the copolymerisation of a dicarboxylic acid and a diamine. They may be of two types, (a) Nylon - 66, made by polymerising hexamethylene diamine and adipic acid (hexanedioic acid). It is used in making bristles for brushes, in textiles and also for making sheets, socks and other hosiery products.

(b) Nylon-6, made as the condensation product of caprolactum ($C_6H_{11}NO$), a heterocyclic compound obtained from cyclohexane. It is used for the manufacture of ropes, type cords, fabrics etc.

O

OCCLUSION : The process in which small amounts of a substance get entrapped in the intestitial sites of other crystalline substances. It may also represent the absorption of a gas by a solid, usually in a catalytic process.

OCHRE : A mineral form of iron (III) oxide. It is red-brown in colour and is used as a pigment.

OCTADECANOIC ACID : $CH_3(CH_2)_{16}COOH$; Stearic acid. A solid saturated carboxylic acid present in fats as its glyceride.

OCTADECENOIC ACID : $CH_3(CH_2)_7CH=CH(CH_2)_7COOH$; Is-9-octadecenoic acid; Oleic acid. It occurs in nature in oils and fats as glycerides.

OCTAH EDRAL COMPLEX : *See complex.*

OCTAHYDRATE : A compound having eight molecules of water of crystallsation per molecule.

OCTANE : C_8H_{18}; A saturated alkane, liquid at room temperature. Octane and its isomers are present in the light oil fraction of crude oil. Iso-octane is considered as arbitray standard of quality of petrol.

OCTANE NUMBER : It is defined as the percentage of iso-octane (by volume) in the mixture of iso-octane and n-heptane which gives the same anti knocking property in a standard engine under identical conditions as that of the fuel under examination.

Iso-octane is given a value of 100 and n-heptane is given a value of zero. Now a fuel 2,2,3-trimethyl pentane is assigned a value 116 and 2,2,3-trimethyl butane is assigned a value of

124, while n-nonane with an octane number of 45 is considered as the worst fuel. In general, more the branching in the carbon chain, more the unsaturation and more the amount of aromatic hydrocarbons, the more will be the octane number. Further, the octane number of a given fuel may be increased by adding small amounts of anti-knocking agents like tetraethyl lead.

Octanoic Acid : $CH_3(CH_2)_6COOH$; Caprylic acid; A saturated, colourless, liquid carboxylic acid.

Octavalent : Describing a valency of eight.

Octaves : *See Newlands law.*

Octet : Whenever the outermost orbit of an atom acquires eight electrons, it becomes stable. Removal of electrons from such a state is not easy. Zero group elements (except helium) have complete octet and are chemically inert. In case of smaller atoms the basis of bonding is explained by this concept. For example in H_2O, CH_4, CO_2 etc., all the concerned atoms have attained such a state of stable configuration by completing the octet. Ions are generally made by the loss or gain of the particular number of electrons so that their octet is complete. For example Cl^-, Mg^{++}, Al^{3+}, have complete octet.

Ohm : ; The S.I. unit of resistance. It is defined as the amount of resistance which is generated when a current of one ampere is passed through a conductor having a potential difference of one volt between the two ends.

Oil : A viscous liquid, organic compound, e.g., petroleum, glycerol etc.

Oil Of Vitriol : An old name for sulphuric (VI) acid.

Oil Sand : Tar sand; Bituminous sand; A sand stone or any other form of porous carbonate rock which has a large concentration of hydrocarbons.

Oleate : A salt or ester of oleic acid.

Oleic Acid : *See alkene.*

Oleic Acid : *See octadecenoic acid.*

OLEUM : $H_2S_2O_7$; Pyosulphuric acid. A colourless fuming liquid formed when sulphur trioxide is dissolved in concentrated sulphuric acid. It is produced by contact process. The oleum formed can then be converted to sulphuric acid of desirable strength by diluting with water.

OLIVINE : $(MgFe)_2SiO_4$; Rock forming silicates ranging from pure magnesium silicate, Mg_2SiO_4, to pure iron silicate (fayalite, Fe_2SiO_4). It is green or brown-green in colour.

ONIUM ION : A postively charged ion formed by the addition of proton on a molecule, e.g., $H_3{}^+O$, hydronium ion; $NH_4{}^+$ ammonium ion; $C_6H_5NH_3{}^+$, anilinium ion.

OPAL : A form of silica. Out of several varieties, some are gemstones of different colours. The common variety is white, but impurities cause yellow, green or red colours. They exhibit opalescence. Black opal has black background against which the colours are displayed.

OPEN CHAIN : A straight or branched chain of which the ends are not joined.

OPEN HEARTH PROCESS : An old method of converting pig iron into steel. In this process pig ion, scrap, etc. are strongly heated in a refractory lined shallow open furnace. Oxygen is passed through during heating to oxidise impurities and to burn off excess carbon. Limestone is added to yield lime which acts as flux.

OPTICAL ACTIVITY : The ability of compounds to rotate the plane of plane polarised light either towards the right or towards the left side. An optically active compound may exist in two forms one of which rotates the plane of plane polarised light to the left side while the other rotates it towards the right side. These forms are called laevorotatory, denoted by l or -ve sign, and dextrorotatory, denoted by d or +ve sign. The two forms are related to each other as object and its non-superimposable mirror image, and are known as enantiomers. When these forms are mixed in equimolar proportions, an optically inactive mixture, called racemic mixture (denoted by dl or ±) is formed.

The cause of optical activity in the compounds is molecular dissymmetry, i.e., absence of plane of symmetry. The most common case of dissymmetry is the presence of a chiral or asymmetric carbon atom (i.e., a carbon atom attached to four different atoms or groups of atoms). Some examples are lactic acid, 2-butanol, tartaric acid, amimo acids. Several naturally occuring compounds are optically active and possess only one enantiomer. For example, d-glucose occurs in nature. The method of synthesising an optically active form in the laboratory is called asymmetric synthessis.

Optical Glass: A variety of glass used in the manufacture of lenses, prisms and other scientific apparatus. Such a glass has high refractive index. It is done so by adding lanthanoid oxides. Crown glass may contain potassium or barium in place of sodium while the glass contains lead oxide.

Optical Isomerism : *See optical activity.*

Optical Rotary Dispersion : The phenomenon by which the effect of wavelength of light on the rotation of plane polarised light is studied. This may be related with the molecular structure.

Optical Rotation : Deflection of the plane of polarised light by an optically active compound. It is of two types – dextro – rotation and laevorotation. *See optical activity also.*

Orbit : The circular definite path of an electron around the nucleus of an atom. The orbits represent the distance of an electron from nucleus and the overall energy of the electron. The higher the orbit number, the higher is the energy of the electron.

Orbital : A region in three dimensional space around the nucleus in which there is a high probability of finding an electron. Orbital can be considered as a cloud of electron charge. The characteristics of an orbital can be ascertained by its quantum numbers. Size and energy are described by the principal quantum number, shape is represented by azimuthal quantum number, while orientation of the orbital is represented by its magnetic quantum number. S-orbital is spherically symmetrical, p-orbital is dumbbell shaped while d-orbitals are double bell shaped.

Molecular orbitals are formed as a result of the linear combination of atomic orbitals. They may be bonding as well as antibonding, depending on the energy and stability. Further, due to difference in the way of overlapping, they may be described as sigma and pi-orbitals.

Hybrid orbitals are obtained by the combination of atomic orbitals of nearly equal energies.

After combination, they rearrange to make new orbitals with the same energy and the same characteristics. They form more stable sigma bonds. Geometry of the molecule depends on the type of hybridisation of its central atom. For example, sp^3 hybrid orbitals are tetrahedrally, arranged, while sp^2 hybridisation results in the planar triangular geometry and sp hybridisation cause linear geometry. The presence of some fully–filled non bonding pairs of electrons (bone pairs) distort the geometry, e.g., ammonia molecule is pyramidal due one fully filled sp^3 hybrid orbital.

Maximum number of orbitals in an atom is given by n^2. Each orbital can accomodate a maximum of two electrons. Further, filling of electrons into various orbitals is based on Aufbau principle and Hund's rule of maximum multiplicity.

ORDER : The summation of concentration terms which determine the rate of the reaction, or sum of the powers to which various concentration terms are raised in the rate law equation. It is not necessarily dependent on the stoichiometry of the balanced chemical equation. Order may be zero (photochemical combination of hydrogen and chlorine), fractional (decomposition of acetaldehyde has order 3/2) or a whole number (combination of hydrogen and iodine has order 2). Order is an experimentally determined quanity.

ORE : A mineral source of an element from which the element can be easily and economically extracted on large scale, e.g, bauxite is an ore of aluminium galena (Pbs) is the ore of lead.

ORE DRESSING : A method for the concentration of ore.

ORGANIC CHEMISTRY : The branch of chemistry concerned with carbon compounds. The latest definition suggests that it is the study of hydrocarbons and their derivatives. By this definition, compounds like carbon dioxide, carbon monoxide, carbonates, bicarbonates, cyanides and cynates of metals may be excluded. Organic compounds consist of a large number of naturally occuring and synthetic compounds of common use.

ORGANOMETALLIC COMPOUND : A compound containing a carbon metal linkage. $(C_2H_5)_4Pb$, tetraethyl lead; RMgX, alkyl magnesium halides, etc. are the important examples. Now, some nonmetals and metalloids are also included this category, like compounds of B, Si etc.

ORTHO : Describing the diatomic molecule made of two nuclei, having the spin in the same direction, e.g., orthohydrogen. In an aromatic compound, ortho describes the neighbouring positions on the ring, e.g., ortho dichlorobenzene has two -Cl groups present on two ricinal positions on the benzene sing.

ORTHO HYDROGEN : *See ortho.*

ORTHO PHOPHORIC ACID : *See phosphoric (V) acid.*

ORTHO PHOSPHOROUS ACID : *See phosphoric acid.*

ORTHO RHOMBIC CRYSTAL : *See crystal system.*

ORTHO SILICATE : *See silicate.*

OSMIRIDIUM : A white naturally occuring alloy of 17-48% osmium and about 49% iridium. It may have small amounts of rhodium, platinum etc. It is used for making electrical contacts and the tips of nibs of pens.

OSMIUM : Os; A hard bluish-white transition metal having atomic number 76 and relative atomic mass 190.2. It belongs to 5d series of the periodic table. It is found associated with platinum and is used in alloys with platinum and iridium.

OSMIUM (IV) OXIDE : OsO_4; Osmium tetraoxide; A yellowish solid, prepared by heating osmium in air. It is used as a catalyst as well as an oxidising agent.

OSMOMETER : *See osmosis.*

OSMOSIS : The flow of solvent molecules through a semipermeable membrane from the solvent to the solution. The process is of fundamental importance in the transport of liquids within the biological systems, as cells membranes are semipermeable in nature. For experiments in the laboratory, chemical synthetic semipermeable membranes, may be made. For example, copper (II) ferrocyanide deposited in the pores of a porous pot by the diffusion of ions from an aqueous copper (II) sulphate and aqueous potassium ferrocyanide acts as a semipermeable membrane.

OSMOTIC PRESSURE : The minimum external presence which is exerted on a solution to prevent the process of osmosis or to prevent the passage of solvent molecules into it, when the solvent and solution are separated by a semipermeable membrane. It can be measured by the Berkley-Hartley method. Osmotic pressure is a colligative property, i.e., independent of the nature of substances used but dependent on the number of moles of particles of solute in a given amount of solution. As such, it is directly proportional to the concentration of the solution (n/v). An equation of state of solutions may thus be obtained as nRT, in which R is a constant and its value is the same as that of the universal gas constant. The equation is based on vant Hoff's laws for solutions, similar to Boyle's law and Charles' law for gases. Osmotic pressure measurements give a very good method for determining molecular masses of the solutes.

OSTWALD'S DILUTION LAW : An expression relating the degree of dissociation of a weak electrolyte with its concentration in the given solution. Degree of dissociation is proportional to the square root of dilution. This law was given by W. Ostwald.

OXIDANT : An oxidising agent is a substance which is capable of gaining electrons and can be easly reduced to oxidise another substance. In rocket fuels, oxidants are used to provide oxygen for combustion, e.g., liquid oxygen, H_2O_2. General oxidants may be halogens, $KMnO_4$, $K_2Cr_2O_7$, MnO_2, HNO_3 etc.

OXIDATION : The process in which an atom, ion or a molecule loses electrons. This process can take place only when there is another substance to gain electrons. Example, Na metal is oxidised in an atmosphere of oxygen or chlorine. A rod of zinc when dipped in a solution of copper (II) sulphate gets oxidised to Zn^{++} ions, while copper (II) ions get reduced to metallic copper. It may also be given in terms of the oxidation number concept. The process in which the oxidation number of an atom or ion is increased is oxidation.

OXIDATION NUMBER : It is the amount of positive or negative charge which may be present on an atom or which appears to be present on it in a compound or ion. Oxidation numbers of substances in their elementary state are always zero while for monoatomic ions, it is equal to the amount of charge. Sum of oxidation numbers of all atoms in a neutral molecule is zero while for an ion it is equal to the net charge. Oxidation numbers of some are given as – H (+1 or -1), O(-2, -1 or +2), F(-1), Cl, Br, I (-1, +1 +3, +5 +7), alkali metals (+1), alkaline earth metals (+2).

OXIDATION-REDUCTION : *See redox.*

OXIDISING AGENT : *See oxidation.*

OXIME : A type of organic compounds containing the C=NOH group. Oximes are formed by reacting hydroxylamine (NH_2OH) with aldehydes or ketones, whether aliphatic or aromatic. For example, when acetaldehyde is treated, acetoxime ($CH_3CH=NOH$) is formed.

Oxo Acid : An acid in which the ionisable hydrogen is linked with oxygen, e.g., sulphuric, nitric, phosphoric acid.

Oxonium Ion : An ion in which positive charge is accomodated on oxygen atom, e.g., hydronium ion H_3O^+; protonated alcohol RO^+H_2 etc.

Oxoprocess : A method of manufacturing aldehydes by passing a mixture of carbon monoxide, hydrogen and alkanes over a cobalt catalyst under particular conditions of temperature and pressure (423K; 100 atmospheres).

Oxyacetylene Welding : Acetylene mixed with oxygen when burnt produces a very high temperature, nearly 3500 K. It enables the welding of all ferrous metals. It may be applied for cutting as well.

Oxygen : O; A colourless, odourless, gaseous element belonging to group VIA and 2nd period of the periodic table. It has atomic number 8 and relative atomic mass 15.9994. It is the most abundant element present in free state in the atmosphere as well as in the combined state in the form of oxides, nitrates, carbonates, sulphates and phosphates of various metals and non metals. Atmospheric oxygen is vital for all living beings, plants as well as animals, to carry out aerobic respiration. Industrially, oxygen may be obtained by the fractional distillation of liquid air. Oxygen is quite a reactive element and may react with a large number of metals at room temperature or on heating to form oxides. It reacts with several non metals also to make acidic oxides.

Oxyhaemoglobin : *See haemoglobin.*

Ozone : O_3; Trioxygen; A blue-coloured gaseous allotrope of oxygen. Poisonous in nature. Prepared by passing oxygen through a silent electric discharge. Ozone is unstable, and decomposes to oxygen. It acts as a good oxidising agent. It is present in the upper layers of the earth's atmosphere and so protects living organisms from the harmful effects of ultraviolet radiations. Large scale use of chlorofluorocarbons may damage the ozone layer and so may prove hazardous. Ozone may be used for the ozonolysis of alkenes which helps in the formation of aldedes and ketones and in the location of double bonds.

Ozonides : *See ozonolysis.*

Ozonolysis : The reaction of ozone and alkene in the presence of solvent, methylene chloride or carbon tetrachloride, and subsequent hydroysis of the ozonide to get carbonyl compounds. The carbonyl compounds may be separated and identified. This identification helps in locating the position of the double bond and establishing the structure of the original alkene.

P

Palladium : Pd; A white ductile metal of d-block, belonging to the 10th group and 5th period of the periodic table. Its atomic number is 46 while relative atomic mass is 106.4. It occurs in some of the ores of copper and nickel. It is used in electrical relays, as a catalyst in the hydrogenation processes and in jewellery. The element was discovered by Woolaston.

Palmitic Acid : *See hexadecanoic acid.*

Pantothenic Acid : A vitamin of the vitamin B-complex group. Its deficiency, in general, does not occur as it is widely present in many foods like cereals, pears, egg yolk, liver etc.

Paper chromatography : A technique of chromatography in which the stationary phase is paper. A spot of mixture is marked on the paper near an edge and it is then suspended vertically in a solvent which rises through the paper by capillary action, carrying the components with it. The movement of the components at different speeds causes the separation. The components on the dried chromatogram are identified by using ultraviolet rays or by spraying certain chemicals like, ninhydrin which gives blue colour with amino acids. Further components can be indentified by the distance they move in a given time (*See R_f values*).

Para : Describing the form of a diatomic molecule in which both the nuclei have opposite spins, e.g., parahydrogen. In benzene, para depicts the diagonally opposite positions (1,4) For example, p-dichlorobenzene is 1,4-dichlorobenzene.

Paraffin : *See alkanes.*

PARAFFIN WAX : A solid obtained from petroleum. It is a mixture of heavier hydrocarbons.

PARAHYDROGEN : *See hydrogen.*

PARAMAGNETISM : *See magnetism.*

PARTIAL IONIC CHARACTER : Due to electronegativity difference, the electrons of a covalent bond are attracted by the element having higher electonegativity. Such a shift causes polarity in the covalent bond. Examples of partial ionic character in the bonds is given by HI, HCl, Li Br. However, the percentage ionic character in these is quite low as compared to that of NaCl or KCl.

PARTIAL PRESSURE : The pressure which a gas would have if it were present alone in the same volume at the same temperature. The sum of partial pressures of all the gases of a mixture is equal to the total pressure. The law relating to this is called Dalton's law of partial pressures.

PARTITION COEFFICIENT : If a solute is in contact with a mixture of two immiscible liquids, it has a different affinity for each phase. The solute gets distributed between two phases in such a way that the ratio of the concentration in one liquid to the other remains constant irrespective of total amount dissolved. Further, the value of this constant for a substance remains constant for two given liquids . This constant is called partition coefficient. The only condition is that the solute should be in the same molecular state in both the liquids.

PASCAL : Pa; The S.I. unit of pressure. 1 Pa = 1 Nm^{-2}. It is related to atmospheric pressure as 1 atmosphere = 101.325×10^3 Pa.

PASCHEN SERIES : A series of lines in the infrared region of the electromagnetic spectrum. These lines correspond to the energy emitted when electrons once excited to higher energy states jump down to the third orbit. Wave number (number of waves per unit distance,) is given as $1/l = (R\ (1/3^2 - 1/n^2)$, the where n is the orbit number and R is the Rydberg's constant.

PASSIVE : Certain metals like iron do not dissolve in concentrated nitric acid due to the formation of a thin oxide layer. Such metals are rendered passive and lose their initial activity. Iron reacts with HCl to liberate H_2 but once it is treated with conc. nitric acid, it does not react with HCl.

Pauli's Exclusion Principle : No two electrons in an atom can have the same set of four quantum numbers. It may also be stated that an orbital can have a maximum of two electrons having opposite spins.

Pearl Ash : Potassium carbonate.

Pearlite : *See steel*

Penicillin : An antibiotic obtained from the fungus *Penicillium notatum*. It is specifically Penicillin-Cr. Penicillins of various kinds are used to treat a variety of infections and are called antibiotics.

Pentahydrate : A crystalline compound having five molecules of water of crystallisation, e.g., $CuSO_4 . 5 H_2O$

Pentane : $CH_3(CH2)_3CH_3$; A straight chain saturated hydrocarbon (alkane).

Pentanoic Acid : $CH_3(CH_2)_3COOH$; Also called valeric acid. It is a colourless liquid acid used in perfumes.

Pentavalent : Having a valency of five.

Pentlandite : Chief ore of nickel.

Pentose : A carbohydrate having five carbon atoms. Simplest carbohydrate, which is not further hydrolysed.

Pentyl Group : CH_3-CH_2-CH_2-CH_2-CH_2 group.

Pepsin : An enzyme which catalyses the hydrolysis of proteins and is secreted by the cells of the stomach.

Peptide : A compound formed by the linkage of two or more alpha amino acids. Peptides have - CONH- linkage. They may be considered as the building units of all proteins.

Per : A prefix denoting the excess of an element, e.g., peroxide, perchlorate, etc.

Perchlorate : Compound having ClO_4^- lions

Perchloric Acid : *See chloric (VII) acid*.

Per disulphuric Acid : *See peroxosulfuric (VI) acid.*

Period : *See periodic table.*

PERIODIC ACID : *See iodic (VII) acid.*

PERIODIC LAW : The law was originally given by Mendeleev. It stated that the properties of the elements are a periodic function of their atomic weights. Mosley, later on modified it as that the properties are a periodic function of the atomic numbers and not the atomic weights. If the elements are arranged in the order of their atomic numbers, elements having similar properties are repeated after a definite gap. This gap of elements may be of 2,8,18 or 32 elements.

PERIODIC TABLE : A table of the elements arranged in the order of their atomic numbers to exhibit similarities in chemical behaviour. Vertical rows in the table are called groups which are 18 in number, while horizontal rows are called period which are 7 in number. There is a regular variation of properties along a period and down a group. For example, metallic property, electopositive nature and size increase down a group, while metallic properties, electropositive character and size decrease while moving from left to right along a period. It may also be divided into four blocks namely s, p, d and f-blocks. Elements of s-block are metals, p-block elements are mainly non-metals, d-block elements transition metals and f-block elements, are called inner transition elements.

PERMALLOYS : A group of alloys which are highly magnetic in nature. Main constituents of these alloys are iron and nickel. They are used in computer memories.

PERMANENT HARDNESS : Hardness caused by the presence of dissolved salts of calcium and magnesium like chlorides and sulphates.

PERMANGANATE : *See manganate (VII)*

PREMUTIT : A substance used to soften the hardness of water. It is a zeolite consisting of a complex chemical compound, sodium aluminate silicate. Calcium and magnesium ions present in hard water are exchanged with sodium ions of permutit. Exhausted permutit can be regenerated by washing it with a concentrated aqueous solution of sodium chloride.

PEROXIDE : Compound containing -O-O- group.

PEROXOSULPHURIC (VI) ACID : H_2SO_5; A white crystalline solid, formed by the action of hydrogen peroxide on sulphuric acid. It is also called caro's acid.

PERSPEX : A trade name for an acrylic resin, polymethylmethacrylate, PMMA

PETA : Prefix denoting 10^{15}.

PETROLEUM : Also called crude oil. It is found below the rocky strata of the earth. It is formed by the decomposition of dead animals and plants due to high temperature and pressure below the earth. It is a mixture of aliphatic hydrocarbons. Crude oil is separated into fractions by fractional distillation. Important fractions are – gasoline (used as motor fuel), diesel oil (used for diesel engines), kerosene oil (used as a fuel and illuminant), petroleum gas (used as fuel), lubricating oil (used as lubricant) and paraffin wax.

PETROLEUM ETHER : A colourless volatile mixture of hydrocarbons, namely pentane and hexan, used as a solvent.

PEWTER : A corrosion resistance alloy of tin and lead (80-90 % tin). Antimony is sometimes added to make it hard, used to make containers.

PH : Negative logarithm of hydrogen in concentration of a solution. pH of pure water at about 25°C is 7. Lower the value of PH, more is the acidic strength of the solution, and higher PH indicates basic solutions. Its range is 0 to 14.

PHASE : One of the physically separable parts of a chemical system. Water at equilibrium with ice is a two phase system. A system having one phase is called a homogenous system while the systems having more phases are called heterogenous.

PHASE DIAGRAM : A graph which relates the solid, liquid and gaesous phases over a range of conditions of temprature and pressure. Lines of the diagram represent the coexistence of two phases in equilibrium while the points on the graph represent three phases existing simultaneously in equilibrium.

PHASE RULE : For a system at equilibrium, the number of phases, components and degrees of freedom are related by a mathematical expression, F + P = C + 2, where F is variance or degrees of freedom, P is number of phases and C is number of components. It was first given by J. W. Gibbs.

PHENOL : C_6H_5OH; A colourless crystalline solid also called carbolic acid. It is prepared by the alkali fusion of sodium. salt of benzene sulphuric acid. It is manufactured by Raschigs process and through the formation of cumene.

PHENOLPHTHALEIN : An aromatic dye used as an indicator for acid-alkali titrations. It gives a light pink or red colour in alkaline medium and becomes colourless in acidic medium. Its pH range is 8-9.6, and so it can be used when a strong base is titrated against a strong acid or against a weak acid. It cannot be used in the reactions in which carbon dioxide gas is evolved.

PHENOLS : Organic compounds which have one or more OH, groups attached directly to the benzene ring. Phenols are acidic in nature due to the resonance stabilisation of phenoxide ion. They are weaker acids than carbonic acid, so do not decompose carbonates and bicarbonates. They may be identified by litmus solution or by neutral ferric chloride solution. Acidic strength of phenols is increased by introducing an electron withdrawing group (like NO_2) in the benzene ring, particularly on its ortho or para position.

PHENYLALANINE : C_6H_5-CH_2-CH-(NH_2) COOH; An alpha amino acid.

PHENYLETHENE : C_6 H_5-CH=CH_2; Styrene; A liquid unsaturated hydrocarbon required to synthesise polystyrene (a polymer).

PHENYL GROUP : $C_{6\ 5}$H-group.

PHENYLHYDRAZONE : A compound prepared by the reaction of an aldehyde or ketone with phenyl hydrazine (C_6 H_5 NH-NH_2). For example, with ethanal it forms - eq CH_3-CH=N-NH=C_6H_5 called acetaldehyde phenyl hydrazone.

PHENYLMETHANOL : $C_6H_5CH_2OH$; Benzyl alcohol; An aromatic primary alcohol. It is made by the disproportionation of benzaldehyde. It is used as a solvent. Group -CH_2OH is an activating group and directs the incoming group to o-, p-positions.

PHILLIPS PROCESS : A process of manufacture of high density polythene. The process takes place at 423K temperature, under 30 atmospheres and in the presence of chromium (III) oxide catalyst supported on silica alumina.

PHOSGENE : $COCl_2$: *See carbonyl chloride.*

PHOSPHAGEN : A substance found in animal tissues which serves as a reserve of chemical energy in the form of high energy phosphate bonds. Some phosphagens include creatine phosphate, arigine phosphate etc.

PHOSPHATES : *See phosphoric (V) acid.*

PHOSPHIDE : A compound of phosphorous with an electropositive element (metal), e.g., Na_3, Ca_3 P_3.

PHOSPHINE : PH_3, A; colourless, toxic gas, slightly soluble in water, having a smell like dead fish. It is prepared by the action of calcium phosphide and water or by the action of yellow phosphorous on a concentrated alkali. It burns in air spontaneously and decomposes into its elements at 723K in the absence of oxygen, but in air it yields phosphorous oxides (like P_4O_6 and P_4O_{10}). It may cause the precipitation of phosphides of metals when treated with metal saltsolutions. It acts as a ligand that can form a coordinate bond with metal ions. Gas mixtures of pure phosphine and the rare gases are used for doping semiconductors. Like its analogue ammonia, it forms salts called phosphonium salts. One or more hydrogen atoms can be replaced by alkyl groups.

PHOSPHIC ACID : H_3PO_2; Hypophosphorous acid; A white crystalline, monobasic acid. It can be prepared by heating yellow phophorous with sodium hydroxide solution. It is a strong reducing agent.

PHOSPHITE : *See phosphoric acid.*

PHOSPHOLIPID : One of a group of lipids having a phosphate group and one or more fatty acids. They may be based on glycerol (glycerophospholipid) or on sphingosine (sphingophospholipid). Phospholipids in general are a major component of cell membranes.

Phosphoric Acid : H_3PO_3; Phosphorous acid, orthophosphorous acid. A colourless, deliquescent dibasic acid prepared by the action of water on phosphorous (III) chloride. It forms two types of anions, $H_2PO_3^-$ and HPO_3^{-2}. The acid and its salts are called phosphonates (formely called phosphites). They are moderate reducing agents. On heating, it decomposes to form phosphin and phosphoric (V) acid.

Phosphonium : PH^{4+} ion

Phosphor : A substance exhibiting luminescence or phosphorescence.

Phosphor Bronze : An alloy of copper, tin and phosphorous (4-10% tin and upto 1 % phosphorous). It is used particularly for marine purposes.

Phosphorescence : A type of luminescence in which the emitted radiations continue for sometime after the source of excitation has been disconnected. The time upto which it continues may not be defined properly.

Phosphoric (V) Acid : H_3PO_4; Orthophosphoric acid; A white, tribasic acid, solid, soluble in water, it is prepared by reacting phosphorous (V) oxide with water or by heating yellow phosphorous with nitric acid. On heating at about 493K, it is converted to pyrophoric acid ($H_4P_2O_7$). When heated to about 593 K, it forms polymeric metaphosphoric acid $(HPO_3)_n$. It forms three series of salts – $H_2PO^-_4$, HPO_4- and PO_4^{-3} -, Phosphoric (V) acid occurs in nature in the form called phosphate rock, $Ca_3(PO_4)_2$. The alkali metal phosphates, except that of lithium, are soluble in water, while other phosphates are insoluble in water. They are used in water softening and as a fertilizer.

Phosphorous Acid : *See phosphoric acid.*

Phosphorous : P; A non metal belonging to group VA and 3rd period having atomic number 15 and relative atomic mass, 30.9738. It occurs in some allotropic forms like white phosphorous, red phophorous and black phosphorous. It occurs in various phosphates from which it is extracted by heating with coke and silica at a high temperature. It is a highly reactive element and forms phosphides and phosphorous (III) and phosphorus (V) compounds. It is an essential element for living organisms and is present in the RNA and DNA molecule in the form of phosphate units. It was discovered by Brandt.

Phosphorous (III) Bromide : PBr_3; A colourless pungent smelling liquid obtained by reacting red phosphorous with bromine. It is an insoluble compound and is prepared in the organic reactions for substituting the -OH group.

Phosphorus (III) Chloride : PCl_3; A colourless, liquid, pungent smelling compound formed by reacting phosphorus with chlorine. It is rapidly hydrolysed by water to phosphoric acid and hydrogen chloride. It is used to replace the -OH group (in the organic compounds) by chlorine.

Phosphorous (V) Chloride : PCl_5; A white solid formed by reacting phosphorous (III) chloride with chlorine. It is hydrolysed by water to phosphoric (V) acid and hydrogen chloride. It is also a chlorinating agent.

Phosphorous (III) Oxide : P_2O_3; A white solid existing as a dimer P_4O_6, having a smell of garlic. It is obtained when phosphorous is ignited in a limited supply of air. In cold water it slowly dissolves to form phosphoric acid. It is easily oxidised in air to phosphorus (V) oxide.

Phosphorus (V) Oxide : P_2O_5; A white solid, existing as its dimer, P_4O_{10}, prepared by burning phosphorus in excess of air or oxygen. It is used as a dehydrating agent.

Phosphorus Oxychloride : $POCl_3$; Also named phosphorus (III) chloride oxide or phosphoryl chloride. It is a colourless liquid prepared by reacting phosphorous (III) chloride with oxygen. It is formed as a byproduct when alcohols are reacted with phosphorus (V) chloride. It forms complexes with metals.

Phosphorous Pentabromide : *See phosphorus (V) bromide.*

Phosphorous Pentachloride : *See phosphorus (V) chloride.*

Phosphorous Pentoxide : *See phosphorus (V) bromide.*

Phosphorous Tribromide : *See phosphorus (III) bromide.*

Phosphrous Trichloride : *See phosphorus (III) chloride*

Phosphorous Trichloride Oxide : *See phosphorous oxychloride.*

Phosphorous Trioxide : *See phosphorous (III) oxide.*

Phosphorul Chloride : *See phosphorous oxychloride.*

Photochemical Reaction : A reaction which proceeds with the help of radiations of the visible or ultraviolet region of the spectrum. Energy required is supplied by the photons of light. Amount of reactants converted to the products is proportional to the quantity of energy absorbed or number of photons released. This gives photochemical efficiency also. Some examples are, photosynthesis in plants, bleaching of coloured substances in the sun, reduction of silver bromide during photography, reaction between hydrogen and chlorine forming hydrogen chloride, halogenation of alkanes, etc.

Photochemistry : The branch of chemistry which deals with the study of the photochemical reactions.

Photochromism : A change of colour taking place in compounds on exposing them to the light.

Photoelectric Effect : The phenomenon of ejection of electrons when the surface of a metal is irradiated by light of suitable frequency. Minimum frequency needed for ejection of electrons is called threshold frequency. Kinetic energy of the photoelectrons emitted depends on the energy or frequency of the incident radiations, while the number of photoelectrons emitted depends on the intensity of radiations. Alkali meatals exhibit this phenomenon most effectively as the ionisation energy of alkali metals is minimum.

Photoelectrochemical Cell : A type of galvanic cell in which one electrode is a photosensitive semiconductor (e.g. gallium arsenide). Current is generated by the radiations incidented on the semiconductor, resulting in the emission and passage of electrons between the electrodes.

Photoelectrons : Electrons emitted from a solid, liquid or gas by the effect of radiation.

Photoionisaiton : The phenomenon of ionisation of atoms or molecules by radiation. It is similar to photoelectric effect but is valid for gases and liquids also.

Photoluminescence : The emission of light from a substance which can absorb energy and get excited. In fact, radiations are emitted when the excited particles return to its ground or lower energy state. It is also called fluorescence, but when it persists even after the removal of the source, it is called phosphorescence.

Photolysis : Breaking of a bond by electromagnetic radiations. Usually these reactions involve free radicals.

Photon : A quantum of electromagnetic radiation. It is the unit of energy dependent on frequency (hf). Photons travel at the speed of light.

Photosensitive Substances : A compound which when exposed to light produces photovoltaic, photoelectric, photoconductive effects. It may be described as a substance which produces a chemical reaction by the photons.

Photosynthesis : The photochemical process by which the green plants convert carbon dioxide of the atmosphere to carbohydrates with the help of a green colouring matter called chlorophyll.

Phthalic Acid : $C_6H_4(COOH)_2$; A colourless, crystalline solid, dibasic, organic aromatic acid. It is 1,2- benzene dicarboxylic acid and is made by the oxidation of napthalene.

Phthalic Anhydride : Phthalic acid on heating strongly loses a molecule of water to form anhydride which is used in making plasticisers and polyster resins.

Physical Chemistry : Branch of chemistry dealing with the relationship between chemical structures and physical properties.

Physiosorption : *See adsorption.*

Pi-Bond : A bond formed as a result of lateral overlapping of half orbitals. It has two lobes one above and one below the sigma bond. It is weaker than the sigma bond.

Pico : p; A prefix denoting 10^{-12}. For example, 10^{-12} m = 1 pm.

PICRATE : A salt or ester of picric acid.

PICRIC ACID : $C_6H_2(OH)(NO_2)_3$–2,4,6,– trinitrophenol. A yellow coloured highly explosive substance prepared by the complete nitration of phenol, but the yield is very low. It is better prepared through a sulphur acid.

PIEZO ELECTRICITY : The generation of an electric charge when some mechanical stress is applied on certain crystals.

PIG IRON : The crude form of iron produced from a blast furnace. In fact, pigs are the minor channels through which molten iron coming out of the blast furnace is passed to cool. It contains impurities of carbon, silicon and some other elements. Percentage of these impurities is more than 3%.

PIGMENT : An insoluble organic or inorganic colouring matter used in paints.

PIPETTE : An apparatus made of a graduated tube, used for measuring and transferring known volumes of liquids and solutions used in volumetric analysis.

PIRSONNITE : $Na_2Co_3CaCO_3.2H_2O$; Mixed hydrated carbonates of sodium and calcium, occuring as a mineral.

PITCH : A dark coloured residue left when useful components are separated from coal tar or wood tar by fractional distillation. It is used as road tar, for waterproofing. etc.

PITCH BLENDE : An ore of uranium, also called uraninite. *See uraninite also.*

PK_a : The negative logarithm of dissociation constant of an acid.

PK_b : The negative logarithm of dissociation constant of a base.

PLANCK'S CONSTANT : h; Named after Max Planck. It is the ratio of energy possesed by a photon to its frequency. Its value is 6.626196 X 10^{-34} Js.

PLANE POLARISATION : Polarisation of electromagnetic radiations in a way that they vibrate only in a single plane.

PLASMA : It is the mixture of ions and electrons.

PLASTER OF PARIS : $CaSO_4 \frac{1}{2}H_2O$; A hemihydrate of Calcium sulphate dihydrate (gypsum), prepared by heating the mineral to about 423 K under control, so that all its water is not lost. It is used in pottery, in moulds, picture frames and for plastering broken parts. The uses are based on its property of setting in excess water. When mixed with water, gypsum crystals are reformed which get interlocked into one another with the increase in volume, giving rise to a very hard mass. During the preparation of plaster of Paris, when the whole of the water is lost it becomes dead burnt plaster and is of no use as such.

PLASTICISER : A substance added to a synthetic resin to make it flexible.

PLASTICS : A term used for synthetic polymers. They are of two types. Thermoplastics can be repeatedly softened by heating and hardened again on cooling, while thermosetting plastics are initially soft but become hard and rigid on heating, and so they cannot be remoulded again and again.

PLATINUM : Pt; A silvery white transition metal having atomic number 78 and relative atomic mass 195.09. It belongs to VI period of the periodic table. It occurs in the combined as well as in native state. In the combined state, it occurs in some nickel and copper ores. It is used in jewellery, laboratory apparatus, alloys and as catalysts. It is neither corroded by atmospheric gases nor by acids.

PLATINUM BLACK : Finely divided platinum obtained as a layer on the surface by evaporating platinum in an inert atmosphere. It is a very good absorbent and so used as electrodes and catalysts.

PLATINUM (II) CHLORIDE : $PtCl_2$; A grey powder prepared by decomposition of platinum (IV) chloride. It is insoluble in water. With concentrated hydrochloric acid, it forms a complex.

PLATINIUM (IV) CHLORIDE : $PtCl_4$; A red hygroscopic solid obtained by heating chloroplatinic acid. It exists as a pentahydrate or as anhydrous salt. Anhydrous state can be obtained by treating the hydrated crystals with concentrated sulphuric acid.

PLATINUM-IRIDIUM : An alloy of platnium and iridium having about 70% platinum. It is hard and more resistant to the attack of chemicals. It is used in jewellery, electrical appliances and for making surgical needles.

PLATINUM METALS : The group of ruthenium, osmium, rhodium, iridium, palladium and platinum metals. They are grouped together due to the similarity of their properties.

PLEXIGLAS : A trade name for acrylic resin, polymethyl methacrylate (PMMA).

PLUMBAGO : Another name for graphite.

PLUMBANE : PbH_4; Lead (IV) hydride; A colourless unstable gas obtained by the action of acids on a mixture of magnesium and lead pellets. Some of its derivatives are quite stable, e.g., trimethylplumbane $(CH_3)_3PbH$.

PLUMBATE : PbO_3^{2-}; A substance formed by the reaction of lead oxide with a strong caustic alkali. In fact, various ions may be formed of which $[(Pb(OH)_6]^{2-}$ is the main.

PLUMBIC : Describing a lead (IV) compound.

PLUMBOUS : Describing a lead (II) compound.

PLUTONIUM : Pu; A silvery, radioactive transuranic metal belonging to the actinides, having atomic number 94 and mass number 244 (of most stable isotope). $P_{u\text{-}239}$ is readily obtained by the neutron bombardment of uranium (natural). $P_{u\text{-}239}$ is a major nuclear fuel used in reactors as well as for nuclear explosives. It was first produced by Seaborg, McMillan and Kennedy.

POISE : A c.g.s. unit of viscosity. 1 poise = 1×10^{-1} Ns m^{-2}

POISON : Catalytic poison is the substance which destroys catalytic activity, e.g., arsenic compounds act as poisons for platinum catalyst.

POLAR : Describing a molecule having permanent dipole moment, e.g., water, ammonia, hydrogen fluoride.

POLAR BOND : A covalent bond formed between the atoms of two different elements so that the bonding electron pair is not equally shared by the two atoms. The shared pair of electron is attracted by the more electronegative atom which acquires a little negative charge and the other gets a little positive charge. A molecule will be polar if it is has a polar bond but when the molecule has several polar bonds, overall polarity depends on the resultant polarity of all bonds.

POLARIMETER : An instrument used to study the activity of optically active compounds. It consists of a light source, polariser, polarimeter tube, analyser and telescope.

POLARISABILITY : The ease with which the electron cloud of an anion is deformed under the influence of a cation. It is the measure of covalent character developed when cations and anions having high charges come close. Fajan's rules are based on this tendency.

POLARISATION : (a) The process of restricting the vibrations of electromagnetic radiations in one direction. (b) The formation of products of the chemical reaction in a voltaic cell in the vicinity of the electrodes, resulting in an increase of resistance to the flow of current. (c) Partial separation of charges in an insulator subjected to an electric field.

POLAR MOLECULE : A molecule which has some net dipole moment as the polar bonds existing in the molecule are not symmetrical.

POLLUTION : Any undesirable, unpleasant and harmful change in the environment due to industrial or social activities of man. Pollution may broadly be of three kinds – air, water and noise pollution. Air is the result of industrial processes involving the burning of fuels, mainly due to oxides of sulphur (causing acid rain), chlorofluorocarbons (causing depletion of ozone layer of the atmosphere), excess carbon dioxide produced by burning fuels and vehicular exhausts (causing the greenhouse effect) and carbon monoxide present in the exhaust gases of vehicles. Photochemical smog caused by the action of light on hydrocarbons and nitrogen oxides emitted by car exhausts is becoming a problem these days. Water pollution is caused by non-biodegradable substances like chlorinated hydrocarbon pesticides (e.g., DDT, BHC and heavy metals like Pb, Cu, Zn). Water supplies get polluted by the leaching of nitrated soil from agricultural land. Increasing level of human excreta is also a cause of water pollution. Some other forms of pollution may be noise pollution caused by aeroplanes, movement of vehicles and industrial noise. Recent pollution problems include the disposal of radioactive water and acid rain as also other non–biodegradable synthetic resins etc.

POLONIUM : Po; A radioactive element of group VI and period VI, having atomic number 84, while the mass number of its most stable isotope is 209. It occurs along with uranium but in very small amounts. It emits alpha radiations. Polonium may be used as a heat and thermoelectric power source particularly in satellites. It was discovered by Madame M. Curie.

POLY : Prefix indicating a polymer, e.g., polythene, polystyrene etc.

POLYAMIDE : A synthetic polymer having -CO-NH- group, e.g., Nylon-66.

POLYATOMIC : Denoting a mole having several atoms. Examples include phosphorous (P_4), sulphur (S_8), benzene ($C_6 H_6$) and methane (CH_4) etc.

POLYCHLOROETHENE : A synthetic polymer made by the union of several chloroethene ($Cl\text{-}H=CH_2$) molecules. It has a variety of uses because of its resistance to colour, chemicals and corrosion. It is also called polyvinyl chloride or PVC.

POLYCRYSTALLINE : Describing a substance made of a large number of minute interlocking crystals.

POLYCYCLIC : Describing a compound having two or more rings.

POLYENE : (Polyethylene or polythene). The simplest synthetic polymer made by the combination of a large number of molecules. It is of two varieties–low density polythene is a soft adhesive while high density polythene is hard.

POLYHYDRIC ALCOHOL : An alcohol which has several -OH groups in its molecule. Glucose, sorbitol, mannitol are some common examples.

POLYMER : A macromolecule made by the combination of a large number of monomers, the repeating units. Polymers may be natural as well as synthetic. Natural polymers may be proteins and polysaccharides, while synthetic polymers are of a large variety like nylon, polythene, bakelite, dacron, PVC etc.

POLYMERISATION : The process in which one or more compounds react to form a macromolecule. Homopolymerisation involves only one monomer (polythene PVC etc.) while copolymerisation involves two or more different kinds of monomer units (nylon, dacron etc.). Addition polymerisation involves the addition of monomers only, while condensation polymerisation involves addition of repeating units or monomers along with the elimination of small molecular species like water, ammonia etc. (e.g., nylon, polypeptides, etc).

POLYMETHANAL : A solid polymer of formaldehyde obtained by the evaporation of an aqueous solution of formaldehyde.

POLYMETHYLEMETHACRYLATE : A thermoplastic acrylic polymer made by polymerising methyl methacrylate ($CH_2=C(CH_3)$ - $COOCH_3$).

POLYMORPHISM : Phenomenon of existence of a substance in several physical forms. It is similar to the concept of allotrophy for elements.

POLYPEPTIDE : A peptide containing several amino acid units. Polypeptides having very high molecular masses are called proteins. Properties of a polypeptide depend on the type and sequence of amino acids concerned.

POLYPROPENE : A synthetic polymer made by the polymerisation of propene. It is a thermoplastic and it resembles polythene in its properties but is stronger and lighter. It is commonly called polypropylene.

POLYSACCHARIDES : Carbohydrates having high molecular mass, formed by the joining of repeating units of monosaccharides. On hydrolysis they break to form disaccharides and monosaccharides. Hydrolysis may be carried out by enzymes or acid. Different types of polysaccharides have different monosaccharide units, e.g., starch is made of glucose units. Common examples of polysaccharides are starch, glycogen, cellulose, etc.

POLYSTYRENE : A synthetic polymer made by the polymerisation of styrene ($C_6H_5CH\text{-}CH_2$). It is used in packing and for insulation.

POLYTETRAFLUOROETHENE : (PTFE); A synthetic polymer made from tetrafluoroethene ($CF_2 = CF_2$). It is temperature resistant and has a low coefficient of friction. It is used for coating cooking utensils (to make them nonstick in nature) and non lubricated bearings.

POLYURETHANE : A synthetic polymer containing the group NH–CO–O. They are made by condensation of isocyanates (-NCO) with alcohols.

POLYVINYL CHLORIDE : *See polychloroethene.*

PORPHYRIN : A group of compounds having a cyclic group of four linked nitrogen-containing rings. They differ in the nature of their side chains. Common examples are chlorophyll, haem groups of haemoglobin, myoglobin, etc.

POTASH : A compound of potassium, like potassium carbonate or potassium hydroxide.

POTASSIUM : *See Alums.*

POTASSIUM : K; A soft silvery alkali metal belonging to group A and the 4th period of the periodic table. It has atomic number 19 and relative atomic mass 39.098. It exists in the combined state as potassium chloride in sea water, as well as in several minerals like caruallite ($KCL\ MgCl_2.\ 6H_2O$), kainite ($MgSO_4\ KCl_3.\ H_2O$), alums, nitre etc. It is a very strong reducing agent and so cannot be extracted by reducing its minerals. It is obtained by the electrolysis of molten chloride. It is an essential element for living beings. Compounds of potassium are very important. It is a highly reactive element and so combines with air, moisture and almost all non metals. It was discovered by Sir Humphry Davy.

POTASSIUM BICARBONATE : *See potassium hydrogen carbonate.*

POTASSIUM BROMIDE : KBr; A white crystalline solid, soluble in water obtained by the action of bromine on caustic potash or by the action of hydrobromic acid on potassium carbonate.

It is used in the maufacture of photographic plates, films, papers etc., and as a prism material for IR spectroscopy.

POTASSIUM CARBONATE : K_2CO_3; Pearl ash. A white powdery deliquescent solid soluble in water. It cannot be prepared by Solvay process because of high solubility of intermediate potassium bicarbonate. It is used in the manufacture of potassium soaps, (soft toilet soaps), shaving creams, hard glass, etc.

POTASSIUM CHLORATE : $KClO_3$; A white solid soluble in water. It is formed along with potassium chloride when chloride gas is passed through hot and concentrated solution of caustic potash. On heating it decomposes to yield oxygen and potassium chloride. It is a powerful oxidising agent and is used in explosives, matches, fireworks etc.

POTASSIUM CHLORIDE : KCl ; A white ionic crystalline solid, soluble in water. It is prepared by neutralising caustic potash or potassium carbonate by hydrochloric acid. It occurs in the minerals sylvine and carballite. It is used as a fertiliser. Its old name is of potash. Its crystals are isomorphous with sodium chloride.

POTASSIUM CHROMATE : K_2CrO_4; A yellow-crystalline solid made by the action of aqueous potassium dichromate and caustic potash. It is highly soluble in water. It is used as an indicator involving Ag^+ ions.

POTASSIUM CYANIDE : KCN; A white crystalline solid, highly soluble and extremely poisonous. It is used as a source for other cyanides and hydrocarbic acids. Aqueous solution of potassium cyanide is alkaline due to anionic hydrolysis. It is used to extract silver and gold.

POTASSIUM DICHROMATE : $K_2Cr_2O_7$; An orange-red coloured crystalline solid, prepared industrially from chromite ore. Soluble in water. It is used as an oxidising agent for a large number of organic compounds. Further, it is used for the estimation of Fe^{++}, I^-, etc., by volumetric analysis.

POTASSIUM HYDRIDE : KH; A white ionic solid prepared by passing hydrogen over heated potassium. It is a strong reducing agent.

Potassium Hydrogen Carbonate : $KHCO_3$; A white solid prepared by passing excess of carbon dioxide through an aqueous solution of potassium chloride or potassium carbonate. In nature it occurs in the mineral calcinite. Its aqueous solution is alkaline due to anionic hydrolysis. On heating, it decomposes to potassium carbonate.

Potassium hydrogen Tartrate : HOOC(CHOH); Cream of tartar; A white crystalline acidic salt used in making baking powders.

Potassium Hydroxide : KOH; Caustic potash; A white solid, highly soluble in water, stongly alkaline and corrosive in action. It is prepared industrially by the electrolysis of an aqueous solution of potassium chloride. It may be prepared by heating potassium carbonate with slaked lime. It resembles sodium hydroxide in all its methods of preparation and properties. It is used as an electrolyte in the nickel-iron battery. It is also used for making soft and liquid soaps.

Potassium Iodate : KIO_3; A white solid formed by adding iodine to a hot and concentrated aqueous solution of caustic potash. It is an oxiding agent. In the presence of a reducing agent and some acid, it liberates iodine.

Potassium Iodide : KI; A white ionic crystalline solid prepared by reacting caustic potash with iodine. It is used in medicines, and is particularly useful for diseases due to iodine deficiency. Dilute solution of KI can reduce potassium permanganate to Mn^{--} salt and copper (II) ions to copper (I) ions.

Potassium Manganate (VII) : $KMnO_4$; Potassium permanganate; A dark purple coloured solid, soluble in water. It is manufactured from pyrolusite ore (MnO_2). It is used in the laboratory particularly for volumetric estimations. It is also used as an antiseptic, bactericide and disinfectant. It is a strong oxidising agent and behaves in different manners in acidic, alkaline and neutral media.

Potassium Monoxide : K_2O; A white ionic solid prepared by heating potassium with potassium nitrate. It dissolves in water to produce potassium hydroxide solution. The reaction is highly vigorous producing a lot of heat.

POTASSIUM NITRATE : KO_3; Saltpetre. A white crystalline solid, soluble in water,. It occurs in nature on the surface of the earth as saltpetre. On heating, it decomposes to nitrite and oxygen. Potassium nitrate is used in gunpowder, fertilisers and for the laboratory preparation of nitric acid.

POTASSIUM NITRITE : KNO_2; A deliquescent, white solid, soluble in water. It is used in the laboratory and industry for the manufacture of azo dyes as it is directly needed for diazotisation.

POTASSIUM PERMANGANATE : *See potassium manganate (VII).*

POTASSIUM SULPHATE : K_2SO_4; A white crystalline solid made by neutralising potassium hydroxide or potassium carbonate with dilute sulphuric acid. It is used as a fertiliser, and for producing alums.

POTASSIUM SULPHIDE : K_2S; A yellowish brown or yellow-red deliquescent solid prepared by passing hydrogen sulphide gas through an aqueous solution of potassium hydroxide. It is manufactured industrially by reducing potassium sulphate with coke at a high temperature in the absence of air. Its aqueous solution is strongly alkaline due to anionic hydrolysis.

POTASSIUM SULPHITE : K_2SO_3; A white crystalline solid soluble in water. It is used as a reducing agent in photography. It may also be used as a food preservative.

POTASSIUM SUPEROXIDE: KO_2; A yellow solid prepared by burning potassium in excess of oxygen. When treated with cold dilute mineral acid, hydrogen peroxide is formed. It is a strong oxidising agent.

POTENTIOMETRIC TITRATION : A titration in which the end point is determined by measuring the potential difference during the titration by immersing an electrode into the solution.

POUNDED : Pdl; The unit of force in 7 ps (foot-pound-second) system. It is equal to 0.138255N.

PRASEDYMIUM : Pr; A soft, silvery, ductile, malleable metal of the lanthanide series (of VI period, 4F-series of elements). Its atomic number is 59 and its relative atomic mass is 140.91. It occurs in the combined state along with the other lanthanides. It is used in mischmetal, a rare earth alloy having 5% Pr. Another alloy with higher percentage of Pr (about 30%) is used as a catalyst in the cracking of crude oil. It was discovered by C.A. Von Welsbach.

PRECIPITATE : A suspension of small particles in a liquid formed by the chemical reaction. Such a substance usually settles at the bottom.

PRECIPITATION : The process of formation of a precipitate in aqueous medium.

PRECURSOR : A compound which leads to the formation of another compound in a series of chemical reactions.

PRESSURE : The force on a surface due to some other surface or a fluid acting at 90° to an unit area of the surface. Pressure = force/area. For gases, pressure is due to the force exerted due to molecular collisions on the unit surface of the wall of the container. Its S.I. unit is pascal, Pa.

PRESSURE GAUGE : A device used to measure pressure. Barometers (Fortin and Aneroid) are used to measure the atmospheric pressure while monometers (closed end and open end) are used to measure the pressure of gases.

PRIMARY ALCOHOL : An alcohol having -CH_2-OH group or -OH group attached to a terminal C-atom.

PRIMARY AMINE : An amine having -NH_2 group linked to an alkyl or aryl group.

PRIMARY CELL : An electrochemical cell in which the chemical reaction producing e.m.f. is not reversible, and so such a cell can not be recharged like a lead accumulator. Examples, dry cell, mercury cell, Weston cell, Leclanche cell, Daniell Cell.

PRIMARY STANDARD : A substance which can be used for making standard solutions. Such a substance should be easy to purify, easy to preserve in pure state, unaffected by air and other atmospheric gases, readily soluble, not easily hydrolysed. Examples include potassium permanganate, potassium, dichromate, oxalic acid, Mohr's salt etc.

PRODUCER GAS : A mixture of carbon monoxide (25-30%), nitrogen (50-55%) and hydrogen (10-15%), obtained by passing air along with a small amount of steam through a thick layer of well-heated coke in a furnace called producer. The producer gas is used as a fuel mainly in the glass industry.

PROGESTERONE : A hormone secreted by the corpus luteum of ovary and by the placenta. It prepares the inner lining of the uterus for implantation of a fertilised egg.

PROGESTOGEN : One of a group of naturally occuring hormones which maintains the normal course of pregnancy, e.g., progesterone. In larger amounts they inhibit the secretion of luteinizing hormone and so check ovulation. Therefore, they are used as a major component of oral contraceptives.

PROLACTIN : Lactogenic, luteotrophic hormone or luteotrophin. A hormone secreted by the pitutary glands. In mammals, it activates mammary glands to produce milk, and corpus luteum of the ovary to secrete the hormone progesterone.

PROLINE : A hydroxypyroline, an amino acid.

PROMETHIUM : Pm; A soft silver radioactive element of the lanthanide series (in VI period). Its atomic number is 61 and relative atomic mass in 145. It does not occur in nature but may be produced by the fission of uranium. It is used as a beta decay power source. It was discovered by J.A. Marinsky, L.E. Gendenin and C.D. Coryell.

PROMOTOR : A substance when added to a catalyst improves its efficiency. For example, alumina or molybdenum promotes the efficiency of finely divided iron used as a catalyst in Haber's process.

PROOF : A measure of ethanol in the alcoholic drinks. Proof spirit contains 42.28% of ethanol by weight or 57.1% by volume. The degree of proof gives the amount of alcohol present, e.g., 60% proof should possess 0.6 x 57.1 or 34.26% ethanol by volume.

PROPANAL : CH_3-CH_2-CHO; Propionaldehyde; A colourless, liquid, aliphatic aldehyde.

PROPANE : C_3H_8; CH_3-CH_2-CH_3; A gaseous saturated hydrocarbon, produced by the cracking of heavier fractions of petroleum. It is used as a fuel.

PROPANE : 1,2,3,-triol;-$CH_2(OH)$ $CH(OH)$ $CH_2(OH)$; Glycerol; A colourless viscous liquid obtained from the mother liquor from which soap has been removed. It is used as a solvent and plasticiser.

PROPONIC ACID : CH_3-CH_2-COOH; Propionic acid; A colourless liquid aliphatic carboxylic acid.

PROPANOL I : CH_3-CH_2-CH_2-OH; n-propyl alcohol; A colourless volatile inflammable liquid, soluble in water.

PROPANOL 2 : CH_3-CH (OH)-CH_3; Iso-propyl alcohol; A colourless, volatile, inflammable liquid, soluble in water obtained by the hydration of propene.

PROPANONE : CH_3COCH_3 ; Acetone; A colourless; liquid, aliphatic ketone used as a solvent, prepared by the oxidation of 2-propanol, which in turn is made by the aerial oxidation of propene. Industrially, it is obtained as a by–product in the manufacture of phenol from cumene.

PROPENE : CH_3-CH$=$$CH_2$; Propylene; A gaseous alkene obtained from heavier fractions by catalytic cracking. It is the starting material of propanol-2 which is needed for the manufacture of acetone, the starting material for polypropylene and the starting material for the manufacture of glycerol by synthesis.

PROPENENITRILE : $CH_2$$=$CH(CN); Acrylonitrile; An organic compound required to produce acrylic polymers.

PROPIONALDEHYDE : *See propanal.*

PROPIONIC ACID : *See propanoic acid.*

PROPYLENE : *See propene*

N-PROPYLE GROUP : CH_3-CH_2-CH_2–group.

ISO-PROPYL GROUP : $(CH_3)_2$CH- group.

PROSTHETIC GROUP : A tightly bound non-peptide inorganic or organic component of a protein. They may be lipids, carbohydrates, ions, phosphate etc.

PROTACTINIUM : Pa; A toxic radioactive metal of the actinide series (5F series) of the periodic table, having atomic number 91 and relative atomic mass 231.036. It occurs in minute quantities in uranium ores. The most stable isotope is Pa-231, with a half life of 3.43×10^4 years. It was discovered by Lise Meitner and Otto Hahn.

PROTEASE : Also called peptidase or proteinase. It is an enzyme which catalyses the breaking of proteins into smaller components called peptides. This process is called proteolysis. It helps in the digestion of protein food to amino acids. Examples are pepsin, typsin, etc.

PROTEIN : An essential compound present in all living matter. It is made of the union of a large number of laevorotatory alpha amino acids into long chains. Properties of a particular type of protein are based on the type and sequence of alpha amino acids constituting it. It is called a primary structure. Secondary structure describes the manner in which the long chains are coiled into helix, while tertiary structure explains the way in which the chain or sheet is arranged in three dimension.

Helix is formed by the H-bonding between N−H and C=O groups. Tertiary structures may have - S-S- (cystine) links also. Proteins may be broadly divided as globular and fibrous proteins. Globular proteins have rounded molecules and are water soluble. They include antibodies, haemoglobin, casein, albumin etc. Fibrous proteins are generally insoluble in water and consist of long coiled strands. They include keratin, collagen, actin, mysion, etc.

PROTEOLYSIS : The enzymatic breaking of proteins by protease. *See protease also.*

PROTON : A fundamental elementary positively charged particle existing in the nucleus of all atoms. Its mass is 1.672614×10^{-27} kg.

PROTONIC ACIDS : An acid which produces hydrogen or hydronium ions (H^+, H_3O^+) in its aqueous solution. Strength of an acid depends on the concentration of H_3O^+ ions.

PROTON NUMBER : Z; Atomic number; The number of protons in the nucleus of an atom or the total positive charge units present in the nucleus. It determines the chemical properties of the element as the atom is electrically neutral and so the electronic configuration depends on the proton number.

PSEUDOAROMATIC : A compound which has conjugate double bonds like that of benzene but does not exhibit aromaticity. They do not obey Huckel's rule. Cyclooctatetrane is a common example.

PSEUDO FIRST ORDER : Describing a reaction that exhibits first order kinetics under special conditions. For example, hydrolysis of sucrose, or hydrolysis of an ester in the presence of a large excess of water, active mass of water remains almost constant throughout and so the rate of the reaction depends only on the concentration of sucrose or ester.

PSEUDOHALOGENS : A group of inorganic compounds which resembles halogens in some reactions. Structurally their molecules are symmetrical like those of halogens. Examples include cyanogen $(CN)_2$ and thiocyanogen $(SCN)_2$.

PTYALIN : An enzyme that helps in the digestion of polysaccharide, starch etc. It is present in the saliva.

PUMICE : A porous volcanic rock having several cavities. It is generally lighter than water and used as an abrasive and for polishing.

PURINE : A nitrogenous cyclic base, insoluble in water which gives rise to several bioloigically important bases like adenine and guanine (constitutents of DNA and RNA).

PYRANOSE : A six membered heterocylic ring (having five carbons and one oxygen atom). Sugars having such a structure are called pyranose.

PYREX : Trade name of a very strong heat, resistant, inert borosilicate glass, commonly used for laboratory glassware and other equipment.

PYRIDINE : C_5H_5N; An organic heterocyclic nitrogenous compound, basic in nature. It is aromatic as the electron could of six pi-electrons is delocalised over the whole of the ring. It is extracted from coal tar. It is used in the preparation of other organic compounds, as a solvent and also as a catalyst for some reactions.

PYRIMIDINE : An organic nitrogenous base having a heterocyclic ring of four C-atoms and two N-atoms. Several derivatives of pyrimidine category include uracil, thymine, cytosine (constitutents of DNA and RNA), etc.

PYRITE: FeS_2; Iron pyrites; A mineral resembling gold in appearance, and is therefore called fool's gold. It is much harder and brittle than gold. It is used to produce sulphur dioxide for the manufacture of sulphuric acid by contact process.

PYRO : Prefix denoting an oxo acid obtained by the dehydration of lower acid, e.g., pyro sulphuric acid, $H_2S_2O_7$.

PYRO ELECTRICITY : The property of certain crystals, like tourmaline, which acquire electric charges on opposite faces when heated.

PYROLYSIS : Chemical decomposition of higher molecules on heating strongly. May be called thermal decomposition.

PYROMETER : An instrument used in the industry to measure very high temperatures.

PYROPHORIC : Igniting in air. Pyrophoric alloys are alloys which give sparks when struck.

PYROPHOSPHORIC ACID: *See phosphoric (V) acid.*

PYROSILICATE : *See silicate.*

PYROSULPHURIC ACID : *See disulphuric (VI) acid.*

PYROXENES : A group of ferromagnesium rock forming silicate minerals developed by metamorphic processes in gneisses, schists and marbles. They are related to the amphiboles from which they differ in cleavage angles. Some of these are orthorhombic pyroxenes, $(Mg,Fe)_2Si_2O_6$. Monoclinic pyroxenes include diopside Ca Mg Si_2 O_6, hedenbergite CA $Fe_{2+}Si_2$ O_6; angite (Ca, Mg, Fe, Ti, $Al)_2$ (Si $Al)_2$ O_6; pigeonite $(MgFe^{++}Ca)$ $(Mg\ Fe^{2+})$ Si_2O_6.

PYRUVIC ACID : $CH_3COCOOH$; 2–oxopropanoic acid; A colourless liquid keto acid. It is formed during glycolysis as an intermediate in the metabolism.

Q

QUARDIVALENT : Valency of four.

QUALITATIVE ANALYSIS : Analysis of a sample in order to identify the components of a sample. It involves simple preliminary tests followed by a scheme of a systematic tests.

QUATITATIVE ANALYSIS : Analysis of a sample in order to determine the concentration of one or more components of a sample, involving volumetric and gravimetric analysis. More modren techniques are also used nowadays.

QUANTISED : Physical quantity that can take certain discrete values. It can have a continuous range of values of E_1, E_2 etc. and not the intermediate energies.

QUANTUM : Amount of energy released or absorbed in a process. Example, quantum of e.m. wave is photon.

QUANTUM EFFICIENCY : The number of molecules easily decomposed for each quantum of energy released.

QUANTUM ELECTRODYNAMICS : Describes how particles and electromagnetic radiations interact with the use of quantum mechanics.

QUANTUM MECHANICS : *See quantum theory.*

QUANTUM NUMBER : An integer or half integer value, which gives the value of quantised physical quantity.

QUANTUM STATES : States of an atom or electron which are specified by unique set of quantum numbers, e.g., an electron in K-shell is specified by four quantum numbers, $n = 1$, $l = 0$, $m = 0$, $n - 1/2$.

QUANTUM THEORY : A mathematical theory introduced by Max Planck, which is based on the view that energy (or some other physical quantity) can be changed only in certain discrete amounts for a given system.

QUANTERING : Common method to get a small representative sample of a solid material.

QUARTET : Group of four closely spaced lines in any spectrum.

QUARTZ : Naturally occuring crystalline form of silica (Si O_2). MP 1600 - 1700°C. Insoluble in all acids except HF.

QUATERNARY AMMONIUM COMPOUND : These are salts, formed by the addition of a proton to an amine. Simplest example is ammonium compound formed from ammonium and an acid. Also called nitrogenous based.

QUENCHING : Method which changes the mechanical properties of metals. In this method hot metals are dipped in oil or a water bath. It increases the hardness of metals.

QUICK LIME : *See calcium oxide.*

QUINHYDRONE ELECTRODE : An electrode used to measure H^+ concentration in any solution for measuring the electrode protential of a substance.

QUINOL : *See 14-cyclohexanediol.*

QUINOLINE : C_9C_7N; BP 238°C. Colourless oily liquid. Hygroscopic in nature. Prepared by Skraup's method.

QUINONE : Benzopuinone.

R

RACEMIC MIXTURE : *See optical activity.*

RACEMISATION : The conversion of an optically active isomer or enantiomer into a racemic mixture, i.e., the mixture of two or more enantiomers in equal proportions to get an optically inactive mixture.

RAD : A unit of absorbed inoising radiation. 1 Rad = 10^{-2}J of energy in Kg of substance.

RADIAN : S.I. unit of plane angle 2π radian is equal to 360°.

RADIATION : Emission of energy from a source as waves (light, sound) or as beams of particles in motion (beta rays, alpha rays, etc.)

RADIATION UNITS : The becquerel (Bq), the S.I. unit of radioactivity is the activity of a nuclide decaying at a rate of one disintegration per sound.

Curie (C) the unit previously used was taken as the activity of 1 gram of Ra-226.

1 C = 3.7×10^{10} Bq.

The gray (Gy), the SI unit of absorbed dose, is the absorbed dose when the energy per unit mass imparted to matter by ionising radiation is 1 joule per kilogram.

1 Rad = 10^{-2} Gy.

In SI system exposure to ionising radiation is expressed in coulombs per Kg, the quantity of x-rays of gamma radiation that produces ion pairs carrying one coulomb of charge of either sign in one Kg of pure dry air.

RADICAL : An atom or group having a free valency on one of its atoms. *See free radical also.*

Radioactive : Describing a substance which exhibits radioactivity.

Radioactive-Dating : A method of measuring the age of materials depending on the radio activity. *See carbon dating for details.*

Radioactive Series : A series of radioactivity nuclides in which each member of the series is formed by the decay of the nuclide prior to it. The series ends with a stable nuclide. Thorium series is initiated by Th-232, actininum series starts with U-235 while uranium series starts with U-238. All these series end at the appropriate stable isotope of lead.

Radioactive Tracing : *See labelling.*

Radiochemistry : The branch of chemistry concerned with the study of radioscopes of various elements. It may involve the preparation of radiosotopes or their occurence, separation, chemical properties and their applications. It may also be termed as nuclear chemistry.

Radiosotopes : A radioactive isotope of an element. Example, C-14 is a radiosotope of carbon. They are very useful and play an important role in research, in the study of mechanism through tracer elements (*see labelling*) and in medicine (for diagonisis as well as treatment).

Radiolysis : Process of carrying on a reaction with the help of high energy radiations (X-rays, uv-rays, alpha particles, beta particles etc.) By such radiations or particles, ions or other species are formed which cause the reaction to proceed.

Radiometric Dating : *See radioactive dating.*

Radiowave : Electromagnetic radiation with large waves, having wavelengths greater than a few millimeters. These waves will have very low energy.

Radium : Ra; A white luminescent radioactive metal of the 2nd group and 7th, period having atomic number 88 and relative atomic mass 226.03. Its most stable isotope is Ra (half life 1602). It is found in pitchblende, an ore of uranium. It is used in luminous paints and in medicine (for diagnosis and treatment - radiotherapy). It was discovered by Marie Curie and Pierre Curie.

Radon : Rn; A colourless monoatomic radioactive gas belonging to the 18th (or Zero) group, and 6th period, having atomic number 86 and relative atomic mass 222. Most of its isotopes are radioactive, most stable being Rn-222 (half–life 3.8days). It is formed by the alpha decay of Ra-226. Radon on decay produces polonium. It was discovered by Ramsey and Gray.

Raney Nicket : A form of nickel obtained by reacting nickel-almunium alloy with sodium hydroxide solution. Aluminium of the alloy dissolved, leaving behind a spongy mass. It is used as a catalyst, particularly for hydrogenation reactions.

Ranksite : A mineral consisting of sodium carbonate, sodium sulphate and potassium chloride. Its formula may be $2Na_2CO_3.9Na_2SO_4.KCL$

Raoult's Law : A relationship between the vapour pressure due to a component above the surface of a solution and its mole fraction in the same solution. For an ideal solution, the higher the mole fraction of a volatile component in the solution, the higher will be the vapour pressure exerted by it. The two are directly proportional to each other. As the vapour pressure to a pure component remains constant at a constant temperature, it may be considered as proportionality constant for the above. Lowering of vapour pressure may be calculated, which is a colligative property and is equal to the mole fraction of a non volatile solute. If the solutions are non ideal they do not obey Raoult's law and there may be negative or positive deviations from it. The law was discovered by the French scientist Francios Raoult.

Rare earths : *See lanthanides.*

Rarefaction : A decrease in the pressure of a gas or liquid. The density decreases by this process.

Rare Gases : *See noble gases.*

Raschig's Process : An industrial method for manufacturing chlorobenzene by a gas phase reaction between benezene, hydrogen chloride and air at about 500 K, and the catalyst for the reaction is copper (II) chloride. The chlorobenezene thus produced may be converted to phenol by using pressure at about 700k.

Rate Constant : k; The constant of proportionality in the rate expression for a chemical reaction, which is equal to the rate of reaction when concentration of every reactant is taken as unity. It is mathematically expressed as rate = k {A} {B} for the reaction A + B → Products. It is independent of the concentrations of the reactions but depends only on temperature. Units of k vary with the order of reaction. For first order reaction unit of k is s^{-1} while for zero order it is mol L^{-1} s^{-1}. It is also known as specific reaction rate of velocity.

Rate Determining Step : The slowest step in a mutli-step reaction. The overall rate of reaction cannot exceed the rate of the slowest step.

Rate Of Reaction : A measure of the amounts reactant consumed or product formed in a chemical reaction in unit time. The factors on which it depends are – nature of the reactants, concentration of reactants, temperature, pressure, radiation, catalyst and the surface area. The units in common use are mol L^{-1} s^{-1} (or mol $dm^{-3}s^{-1}$).

Rationalised Units : A system units in which the equations have been related to the geometry of the system. S.I. units are rationalised. The formulae involving circular geometry contain a factor 2π, while those involving spherical geometry contain a factor 4π. C.g.s. system is not a rationalised system.

Rayon : A textile made artifically from cellulose. There are two types of rayon, both made from wood pulp. Viscose rayon is prepared by dissolving the pulp in carbon disulphide and caustic soda. Such a solution is forced through a fine nozzle into an acid bath to get fibres. On the other hand, acetate rayon is prepared by dissolving celluose acetate (obtained by treating wood pulp with acetic acid) in an organic solvent and forcing the solution through a nozzle in air. The organic solvent used evaporates in air leaving behind the fibres.

Reactant : A substance which is taking part in a chemical process.

Reaction : A process in which the old bonds of reacting molecules are broken and new bonds of products are formed.

Reactors : The various types of vessels (tanks, towers, tubes, etc.) in which chemical reactions are carried out. Special types of reactors, called atomic reactors carry out nuclear reactions.

Reagent : A term used for common laboratory chemicals required for carrying out experiments based on reactions. Example, hydrochloride acid, barium chloride, silver nitrate, nitric acid, caustic soda, etc.

Realgar : A mineral red in colour containing arsenic (III) sulphide.

Real Gas : A substance which exists as a gas with properties different than that of an ideal gas. A real gas may approach ideal gas behaviour at high temperature and low pressure. Its molecules have a finite size and attract one another.

Rearrangement : A type of chemical reaction in which the atoms and groups of a compound rearrange to form some new compound. During rearrangement the pi-electron pairs and other pairs of electrons may also interchange their positions.

Reciprocal Proportion's (Law Of) : Also called law of equivalent proportions, which states that if two elements, A and B, form compounds with a third element, C separately, then the ratio in which they combine with the fixed mass of C is the same or a simple multiple of the ratio of mass they themselves combine with each other. For example, carbon and hydrogen combine with a fixed mass of oxygen to form carbon dioxide and water in a ratio C : C :: 6 : 2 which is the same when carbon and hydrogen combine to form methane (12 : 4).

Recombination Process : The process in which a neutral atom or molecule is formed by the combination between a negative ion or electron and cation. Usually, in such a process the neutral species formed are in the excited state and so they may disintegrate with the emission of electromagnetic radiations.

Recrystallisation : Crystallising a compound several times to get its pure sample.

Rectification : A method of purifying a compound by fractional distillation.

Rectified Spirit : An azeotropic mixture of 95.6% ethanal in water (by volume).

Red Lead : *See dilead (II) lead (IV) oxide.*

Redox : Relating to the process of oxidation and reduction which proceed stimultaneously. If one of the substances is oxidised the other one must be reduced. A simple reaction between hydrogen and chlorine forming hydrogen chloride is its example. In this the oxidation number of hydrogen is increased from zero to +1 while that of chlorine is decreased from zero to -1.

Reducing Agent : A substance which is capable of reducing another substance. The reducing agent gets oxidised itself. For example, ferrous sulphate acts as a reducing agent and reduces potassium permanganate to manganese (II) state in acidic medium.

Reductant : Reducing agent is also called reductant.

Reduction : The process of addition of hydrogen on an electro–positive atom or loss of oxygen on an electronegative atom. On the basis of the electronic concept, it is the process of gain of electrons or the process in which oxidation number of one of the atoms decreases. For example, $Zn + Cu^{++} \rightarrow Zn^{++} + Cu$, in this reaction Cu^{++} ions are reduced and Zn is the reducing agent.

Reduction Potential : *See electrode potential.*

Reference Electrode : Standard electrode. Primary reference electrode is the normal hydrogen electrode while secondary reference electrodes are calomel electrode, silver-chloride electrode.

Refining : Purification of a compound by removing impurities or extracting a compound from a mixture.

Refluxing : The process of boiling a liquid in such a way that the condensed liquid moves back into the vessel. A reflux air or water condenser may be used for this purpose. It may allow the liquid to boil for a pretty long time without loss. Refluxing is used in a large number of organic reactions, particularly those which require heating for a longer time in the presence of a solvent.

REFORMING : The conversion of a long straight chain hydrocarbon into a branched chain hydrocarbon, either by strongly heating or by certain catalytic reactions. It helps in raising the octane number of a given sample of petrol. Further, benzene can be manufactured from n-hexane in the presence of a catalyst. Steam reforming is another process by which methane is changed into a mixture of carbon monoxide and hydrogen in the presence of a catalyst, Ni, at high temperature of about 1173 K. Methylcyclohexane on reforming forms toluene.

REFORMATSKY REACTION : A reaction by which α, β - unsaturated esters are obtained from carbonyl compounds. Actual reaction involves the reaction of a carbonyl compound in the presence of zinc powder to form hydrogen ester. Such an ester on heating produces unsaturated ester.

REFRACTORY : A compound which has a very high melting point Usually, they are inorganic oxides and are used in furnaces.

REFRIGERATION : A process for acquiring low temperatue with the help of compressors. Freons are used as refrigerants these days. They are supposed to damage the ozone layer of the atmosphere and so their use is being restricted.

REGNAULT'S METHOD : A method to determine the density of gases. It is done by weighing the bulb with the gas at a definite pressure and temperature when it is completely evacuated. It can be applied to obtain the molecular weight of the gas.

RELATIVE ATOMIC MASS : The average relative mass of an atom of the element as compared to 1/12th of the mass of a C-12 atom taken as one a.m.u. For example, relative atomic mass of chlorine is 35.5 which means an atom of chlorine (on the average) is 35.5 times heavier than 1/12 th of the mass of C-12 atom.

Relative Density : The ratio of the density of a given substance to the density of some standard substance. Usually the density is measured with reference to water at 4°C. It is also called specific gravity.

Relative Molecular Mass : The average relative mass of a molecule of the substance as compared to 1/12th of the mass of C-12 taken as 1 a.m.u. For example, the molecular mass of glucose is 180 a.m.u.

Relaxation : The process by which an excited substance loses energy and comes down to the lower or even ground energy state.

Rem : (Radiation equivalent Man) ; A unit for measuring the effect of radiation on the body. One rem is defined as the radiation absorbed to one rad. The biological effects depend on the radiation and the energy associated with it.

Renin : An enzyme responsible for the coagulation of milk in the stomach. It converts protein to its insoluble state casein. The process helps provide more time for the digestion of milk in the stomach.

Resin : A natural or synthetic polymer. Natural resins are secreted by several trees and work as gums and mucilages, while synthetic resins are used in making plastics.

Resolution : The method of separating a racemic mixture into its optically active enantiomers. They cannot separated by physical methods as their physical properties like boiling point, melting point, solubility etc. are the same. Some racemic mixtures are resolved by the mechanical separation method, particularly when the crystals of two enantionmers are physically different. In most of the cases they are separated by converting them to diastereomers as their physical properties are different. In some cases, biological methods are used but in these only one of the enantiomer is separated.

Resonance : The properties of many compounds cannot be properly explained by a single structure involving simple covalent bonds. The actual bonding in the molecule can be regarded as a hybrid of two or more contributory forms called canonical or resonance forms. For example, benzene is the resonance hybrid, and is supposed to have intermediate character of bonds, which is confirmed by the bond distance also. C - C bond distance in benzene is 1,397 A° as compared to C-C single bond distance of 15.4 A° and C = C double bond distance of 1.34. A°. Two or more forms which are separated by double headed arrows are only the theoretical forms which explain the behaviour of the molecule. The resonance hybrid is never an equilibrium mixture of the two or more forms. Resonance hybrid is more stable than any of its forms. The difference of energy between that of the actual molecule and the most stable canonical form is called resonance energy. The higher the resonance energy, the higher is the stability of the compound. Most of the properties of aromatic compounds are explained with its help.

Resonance Effect : The interchange of positions of pi-electron pairs and lone pairs of electrons with one another, keeping the arrangements of atoms unchanged. If electrons move away from the group it is called +M effect as in chlorobenzene while the effect is called -M, when electron density moves towards the group in the form of pi-electron pairs, as in nitrobenezene. It is called mesomeric effect. Properties can be expressed with its help.

Resonance Energy : *See resonance.*

Retinol : Another name for vitamin A.

Retort : An apparatus made of glass or metal, having a bulb and a long narrow neck, used for several type of reactions, including distillation.

Reverberatory Furnace : A furnace used for metallurgical operations. It consists of a shallow hearth on which the concentrated ore is heated by flames that pass over it and the radiations are deflected from the roof. It is generally used for calcination, roasting and smelting of concentrated ore.

Reverse Osmosis : By the application of pressure, a solvent may be separated from a solution by a semipermeable membrane. The principle is applied for obtaining pure water by desalination under a high pressure of about 25 atmosphere. Water is pressed to move out of brine through the membrane.

REVERSIBLE PROCESS : A process in which the variable properties which define the state of the system can be made to change in such a way that they pass thorugh the same values in reverse order when the process is reversed. In thermodynamic reversibility all the steps should be retraced during the reverse process. This is generally a very slow process. Naturally occuring processes are irreversible and they do not comply with the above conditions of reversbility.

REVERSIBLE REACTION : A chemical reaction whose direction can be reversed and so may proceed in both the forward and backward directions. For example, the reaction -
$SO_2 + O_2 \rightarrow 2SO_3$ is a reversible reaction. Such reactions generally do not go to completion, and an equilibrium is attained between the reactants and the products.

R_f VALUE : A term used in chromotography, denoting the ratio of the distance travelled by the solvent front to the distance travelled by the sample. It is characteristic of a particular compound. With the help of R_f values, substances can be identified in the chromatogram obtained in case of paper chromatography or thin layer chromatography.

RHENIUM : Re; A silvery white, rare transition metal belonging to group VII B and 6th period of the periodic table. Its atomic number is 75 and relative atomic mass is 186.2. It occurs with molybdenum in nature and is obtained as a by product in refining molybdenum. It is used in making alloys (Mo-Re alloys are superconducting) and is used as catalyst.

RHEOLOGY : The branch of study dealing with the deformation and flow of matter.

RHODIUM : Rh; A silvery white, transition metal belonging to group VII and the 5th period of the periodic table. Its atomic number is 45 and relative atomic mass is 102.9. It occurs with platinum and is used in protective finishes, alloys with platinum, and in mirrors formed by electrolytic deposition. It is highly resistant to acids and chemicals. It was discovered by W.H.Wollaston.

RHOMBIC CRYSTAL : *See crystal system.*

RIBOFLAVIM : *See vitamin.*

RIBONUCLEIC ACID : *See nucleic acids.*

RIBOSE : C_5 H_{10} O_5; A monosaccharide pentose which is an important component of RNA. Its derivtive deoxyribose ($C_5H_{10}O_4$) is a constituent of DNA.

RIBULOSE : $C_5H_{10}O_5$; A monosaccharide pentose, an isomer of ribose and an important intermediate in photosynthesis in plants.

RING : A closed chain of atoms in a molecule, e.g., benzene cyclohexane, cyclopentane, pyridine etc. Fused ring compounds have two or more rings joined to one another in such a way that they share two atoms, e.g., naphthalene, anthracene etc.

RING CLOSURE : A reaction in which one part of an open chain reacts with another part so that a ring is formed, e.g., conversion of n-hexane to cyclohexane by its dehydrogenation.

ROCK SALT : A transparent naturally occuring mineral form of sodium chloride.

ROENTGEN : R; A unit of radiation based on its ionising effect on air. One roentgen can induce 2.58×10^{-4} coulomb of charge in one kg of air.

ROSE'S METAL : An alloy made of 50% bismuth, 25-28% lead and 22-25% tin. Due to its low melting point, it is used in fire protective devices.

ROTARY DRYERS : Certain devices used in the chemical industry for mixing and drying operations. They may be applied for continuous process of drying or for drying in batches.

RUBBER : A naturally occuring polymer. It is a polymer of isoprene (2-methyl-1,3-butadience) and is made from the latex of rubber tree. Synthetic rubbers are made by the polymerisation of chloroprene (neoprene rubber), bitadiene and styrene (Buna-S rubber), bubadiere and acrylonitrile (Buna–N rubber) etc.

RUBIDIUM : Rb; A soft silvery alkali metal belonging to group IA and 5th period of the periodic table. Its atomic number is 37, while its relative atomic mass is 85.47. It is present with silicate minerals of certain metals including lepidolite. It is used in vacuum tubes, photovoltaic cells and for rubidium strontium dating. It was discovered by R.W.Bunsen and G.R.Kirchoff.

Rubidium Strontium Dating : A method of finding the age of a geological specimen based on the decay of Rb-87 to Sr-87 (more stable isotope).

Ruby : The transparent red variety of corundum containing a little of chromiun. They are precious gemstones, even more valuable than diamonds. Industrially they are used in lasers, watches and other sensitive and precision devices.

Rusting : The corrosion of iron in moist air to form a red brown layer on the surface. Chemically, it is $Fe_2O_3 . H_2O$ (hydrated iron (III) oxide). The formulation of rust can be illustrated by electrochemical theory. It is accelerated by impurities in iron, cuts, grooves and wrinkles on the surface of iron, presence of salts in water, humidity in air, etc. It may be prevented by the protective coating of oil, paint, varnish an antirust solution like ammonium phosphate, galvanisation, electroplating of nickel, chromium etc.

Ruthenium : Ru; A white, hard, transition metal of group VIII and 5th period. It has atomic number 44 and relative atomic mass 101.07. It occurs in nature alongwith platinum. It is used in making alloys with platinum, which are used in electrical contacts and as catalysts. It is also used in jewellery. It was first isolated by K.K. Klaus.

Rutherfordium : Another name of Kurchatovium, element no. 104. *See kurchatovium.*

Rutile : A mineral form of titanium (IV) oxide.

Rydberg's Constant : R; A constant with whose help wavelengths of various spectral lines are calculated. Its value is 109678 cm^{-1} for hydrogen. It is equal to $R = 2\pi^2 mz^2 e^4/h^3 C$.

S

SACCHARIDE : *See sugar*

SACCHARIC ACID : $C_6H_{10}O_8$; A colourless, crystalline solid, soluble in water, obtained by the oxidation of glucose.

SACCHARIN : $C_7H_5NO_3S$; White crystalline solid, prepared by the oxidation of toluene–O–sulphonamide. Its sodium salt is used as a non–sugar sweetening agent.

SACHRE REACTION : A process for the industrial preparation of acetylene from natural gas (methane). A part of methane is burnt to raise the temperature (to about 1775 K) of the furnace and a part of it is changed to acetylene and hydrogen at this high temperature.

SACRIFICIAL PROTECTION : A method of protecting the surface of a metal against corrosion. Example, zinc may protect iron at its own cost (galvanisation), Mg pieces linked with iron pipes buried underground protect these pipes (cathodic protection).

SALICYLALDEHYDE : $C_7H_6O_2$; 2-hydroxy benzaldehyde; An oily liquid prepared by the action of chloroform or phenol in the pressence of a caustic alkali (Reimer–Tiemann reaction).

SALICYLIC ACID : $C_7H_6O_3$; 2–hydroxybenzoic acid; A white powdery solid, insoluble in water which occurs as methyl salicylate (oil of wintergreen); prepared by the action of carbon tetrachloride on phenol) in the presence of a caustic alkali (Reimer-Tiemann reaction). Industrially, it is made by the action of sodium phenoxide and carbon dioxide under high pressure (Kolbe's reaction) used in medicines and dyes.

SALINE : Containing a salt like NaCl, KCl, etc.

SALOL : $C_6H_{10}O_3$; Phenyl salicylate, prepared by the action of sodium phenoxide and sodium salicylate in the presence of phosphoryl chloride used as an antiseptic.

SALT : A substance made by the neutralisation reaction between an acid and an alkali. It has two components, a cation (from an alkali, metallic) and an anion (from an acid), e.g., KCl, $NaNO_3$, $Al_2(SO_4)_3$ etc.

SALT BRIDGE : An electrical contact between two half cells of an electro-chemical cell. It is a U-shaped glass tube containing potassium chloride solution in water or in agar. It maintains overall electrial neutrality.

SALT CAKE : Another name for sodium sulphate.

SALT HYDRATE : A salt in which the metal ions are enveloped by a definite number of water molecules. They are bound to the metal ion by ion-dipole interactions. Salts having large number of water molecules show efflorescence.

SALTING OUT : A term used for separating soaps after the saponification of oils and fats. For this, a saturated solution of NaCl is added. It may be used for purification of NaCl by passing HCL gas through saturated solution of impure sodium chloride.

SALTPETRE : *See potassium nitrate.*

SAMARIUM : Sm; A silvery soft metal of the lantharide series of the 6th period of the periodic table. Its atomic number is 62 and relative atomic mass is 150.35. It occurs along with other lanthanides in monazite and bastnatite. It is used in permanent magnets, optical glasses and in the nuclear rections. It was discovered by Francois Lecor de Boisbaudran.

SAND : Obtained by the weathering of rocks. Sand is the mineral containing particles of quartz.

SANDMEYER'S REACTIONS : A method for preparing aryl chlorides and bromides by the action of diazonium salt with copper (I) salt dissolved in the corresponding halogen acid.

SANDWICH COMPOUND : A complex formed by a transition metal ion sandwiched between the rings, e.g., ferrocene.

SAPONIFICATION : Alkaline hydrolysis of an ester. Saponification of oils and fats (glycerides of higher carboxylic acids) forms soaps.

SAPPHIRE : The blue coloured variety of corundum (aluminium oxide). They are obtained from igneous and metamorphic rocks. They are used as gemstones.

SATURATED COMPOUND : An organic compound which has no double or triple bond. Such a compound undergoes substitution reactions.

SATURATED SOLUTION : A solution which contains the maximum amount of solute dissolved at a given temperature. An equilibrium exists between the solid solute and the solution.

S-BLOCK ELEMENTS : Elements of the 1st and 2nd group of the periodic table. They are alkali metals (having configuration ns^1) and alkaline earth metals (having configuration ns^2). They are all metals, form ionic compounds, form alkaline oxides and hydroxides, etc.

SCANDIUM : Sc; A silvery element, the first member of the first transition series, belonging to group IIIB and 4th period of the periodic table. Its atomic number is 21 and relative atomic mass is 44.956. It is used in high intensity lights and in electronic devices. Even before its discovery, Mendleev predicted and named it ekaboron. Later on Nelson discovered its existence as its oxide (scandia).

SCHEELITE : A mineral form of calcium tungstate, $CaWO_4$. It is used to extract tungsten.

SCHIFF'S BASE : A compound formed by reacting an aldehyde or a ketone with primary amine, e.g., acetaldehyde and aniline produed ethylidine aniline ($CH_3\,CH{=}NC_6H_5$). They are also called Anils.

SCHIFF'S REAGENT : An aqueous solution of magenta dye (Rosaniline hydrochloride) decolourised by passing sulphur dioxide. On addition of a reducing agent like an aldehyde, the colour is restored.

SCHORITE : A mineral form of potassium sulphate.

SCHOTTEN BAUMANN REACTION: A process of benzoylation of a phenol or aniline by benzoyl chloride in the presence of a base. By this reaction aromatic esters like phenyl benzoate or substituted amides like N-phenyl benzamide are prepared.

SCHOTTKY DEFECT : A point defect caused by the loss of appropriate number of anions and cations. It is a stoichiometric defect leading to the lowering of density.

SCHROEDINGER'S EQUATION : A quantum mechanical equation describing the wave function of a particle in there dimensional space. When applied to an atom, it leads to three quantum numbers, namely, principle (n), azimuthal (l) and magnetic quantum (m)

$$\frac{\delta^2\Psi}{\delta x^2} + \frac{\delta^2\Psi}{\delta y^2} + \frac{\delta^2\Psi}{\delta z^2} + \frac{8\Pi^2 M}{h^2}(E - V)\,\Psi = 0$$

In this equation, h is Plank's constant, m is the mass of electron, E is the total energy, V is the potential energy, Ψ is wave fucntion.

SCRUBBER : A part of a chemical plant that removes impurities from a gas by washing with water showers.

SECOND : s; The S.I. unit of time.

SECONDARY ALCOHOLS : Alcohols having CH(OH) group, e.g., 2–propanol.

SECONDARY AMINE : An amine having NH group, e.g., N-methyl methanamine $(CH_3)_2$ NH.

SECONDARY CELL : A galvanic cell which is reversible in nature and so may be recharged by passing current through it in the opposite direction, from an external source, e.g., lead accumlator.

SECOND ORDER REACTION : A reaction in which the actual rate of the reactions is proportional to the product of concentration of two reactants, i.e., Rate α [A][B].

Alkaline hydrolysis of ester, decomposition of hydrogen iodide are its examples. Units of rate constant for this are L mol^{-1} s^{-1}. Half–life period of second order is dependent on the original concentration.

SEDIMENTATION : The setting of a suspension. It may take place as such or with the help of an electrolyte, e.g., dirt and dust may be precipitated down by adding alum.

SEED : A small amount of crystals added to a solution to hasten the process of precipitation. For example, bauxite is purified by Bayer's process, by dissolving in caustic soda, and then the solution is agitated to get the precipitate of aluminium hydroxide. This precipitation is hastened by adding a little precipitate of $Al(OH)_3$ from the previous lot, which acts as seed.

SELENIDES : Binary compounds of selenium. Selenides of non metals are covalent, e.g., H_2Se, while those of metals are ionic having Se^{--} ions.

SELENIUM : Se; A grey coloured metalloid element of group VI A and 4th period. Its atomic number is 34 and relative atomic mass is 78.96. It exists in several allotropic forms. The commonly occurring grey form is light sensitive and is used in photocells, solar cells and in some glasses. It is a semiconductor. Chemically, it resembles sulphur. It was discovered by J.J. Berzelius.

SEMICARBAZONES : Organic compounds formed by the action of semicarbazide ($NH_2NHCONH_2$).

SEMICONDUCTOR : A substance which conducts electricity under certain conditions. Conductance in these substances is due to small amounts of impurities. The conductance in semiconductors increases with rise of temperature. They are of two types – n-type and p-type semiconductors. They are used in electronic equipment.

SEMIPERMEABLE MEMBRANE : A membrane which permits the passages of solvent molecules only through it. Such membranes are used for osmosis. Synthetic semipermeable membranes are supported on a porous material. Copper ferrocyanide, when deposited in the pores of a porous pot, makes it semipermeable. In general, egg membrane and cell walls act as semipermeable membranes.

SEMIPOLAR BOND : A bond having slight polarity, e.g., a dative bond.

SEPTIVALENT : Another name for heptavalent, i.e., having valency of seven.

SEQUESTRATION : Complex formation to restrict the general activity of particular type of ions. Ethylene diamine, EDTA, etc. may be used as chelating agents for this purpose.

SERINE : $CH_2OHCH(NH_2)COOH$; An alpha amino acid.

SERPENTINE : $Mg_3Si_2O_5(OH)_4$; Hydrated magnesium silicate minerals. It occurs mainly in two forms, chyrsotile and antigorite.

SERPENTINITE : A rock consisting mainly of serpentine, used as an ornamental stone.

SESQUI : Indicating the ratio of 2:3 in compounds. A sesquioxide may be written as M_2O_3.

SEXIVALENT : Represents a valency of six.

SHELL : A circular path around the nucleus in which electrons can move. They are denoted by letters K,L, M, N. As long as an electron is present in a particular shell, its energy remains unchanged. Farther is the shell from the nucleus, the more will be its energy.

SHERADIZING : The process by which the surface of iron article may be coated with zinc when heated at a temperature of about 645 K. Two metals iron and zinc combine to make internal layers and pure zinc makes the external layer. It was given by Sherad Cowper-Coles.

SHORT PERIODS : *See period.*

SIAL : The rocks forming the earth's crust. These are rich in silica (SiO_2) and aluminium (Al) as shown by the name.

SIDE CHAIN : *See chain.*

SIDE REACTION : A chemical reaction that takes place to a limited extent along with the main reaction.

SIDERITE : $FeCO_3$; A brown coloured mineral form of iron (II) carbonate.

SIEMENS : S, The S.I. unit of conductance I $S = 1\ Ohm^{-1}$

SIGMA BOND : A bond formed by the axial overlapping of orbitals of two atoms only. One sigma bond can be formed between two atoms.

Silanes : A group of silicon hydrides like SiH_4, Si_2H_6, Si_3H_8 etc. They ignite spontaneously in air.

Silica : *See silicon (IV) oxide.*

Silica Gel : A gel made by coagulating sodium silicate sol. It is used in desiccators, balances, etc., to remove moisture. Further, it is used as an adsorbent in chromatography.

Silicanese : *See silanes.*

Silicates : A large number of compounds containing discrete SiO_4^{4-} tetrahedral units or the polymeric SiO_4^{4-} units linked in chains or rings etc. Zircon ($ZrSiO_4$) is the example of orthosilicates having discrete units. Pyrosilicates ($Si_2O_6^{6-}$) are formed by the sharing of one oxygen atom between two such tetrahedral. $Sc_2Si_2O_7$ is a common example. When linking is extended, chains are formed.

An example of cyclic silicate is beryl, $BCl_3\ Al_2\ Si_6\ O_{18}$. When three O atoms of each tetrahedron are shared, the silicate formed is called sheet silicate, e.g., talc, $Mg_3\ (OH)_2\ (Si_4O_{10})$ and kaolin $Al_{12}\ (OH)_4\ (Si_2O_5)$. A three dimensional silicate has the general formula $(SiO_2)_n$ as all the four O-atoms shared. Its example may be silica, SiO_2.

Silicide : A compound of silicon with some more electropositive elements. They produce silanes on reacting with acids.

Silicon : Si, A hard, grey coloured metalloid element of group IVA and 3rd period of the periodic table. Its atomic number is 14 and relative atomic mass is 28.086. It is the second most abundant element in the earth's crust (about 25.6% by weight).· It is less reactive than carbon. Large number of silicates are known (*See silicates*). It forms a large number of organic silicon compounds called siloxanes. The element was first identified by Lavoisier.

Silicon Carbide : SiC; Carborundum. *See carborundum for details.*

Silicon Dioxide : *See silicon (IV) oxide.*

Silicones : Polymeric compounds of silicon having R_2SiO units where R may be an alkyl or phenyl group. Silicones are used as lubricants, water repellents, in waxes and varnishes.

Silicone rubbers are better then natural rubber due to its resistance to heat, acids and chemicals.

Silicon (IV) Oxide : SiO_2; Silicon dioxide; Silica; A hard crystalline compound. It is a three dimensional network structurally. It is used in the manufacture of glass and as flux in some metallurgical operations.

Silver : Ag; A coinage metal belonging to d-block of the periodic table in group IB and 5th period. Its atomic number is 47 and relative atomic mass is 107.87. It occurs in the combined state as silver glance (Ag_2S) and horn silver (AgCl). It is used to make coinage alloys jewellery, cutlery, decoration pieces and in photography.

Silver (I) Bromide : AgBr; A pale yellow solid obtained as a precipitate by mixing any soluble bromide with silver (I) nitrate solution. It is soluble in concentrated solution of ammonia. It is a photosensitive substance and so is used in making photographic films and papers.

Silver (I) Chloride : AgCl; It occurs in nature as horn silver. It dissolves in ammonia solution. It may be used in photography. With ammonia it forms amines, e.g., $AgCl_2\ NH_3$, $AgCl_3\ NH_2$.

Silver (I) Iodide : AgI; It occurs in nature as iodoargyrite. It is a yellow coloured solid insoluble in ammonia solution.

Silver Mirror Test : A test for aldehyde group. A little of the sample is warmed with ammonical silver nitrate (Tollen's reagent). Silver metal is deposited on the inner glass walls of the container.

Silver (I) Nitrate : $AgNO_3$; Lunar caustic; A white crystalline solid, soluble in water. On heating, it decomposes to yield silver. It is used for the treatment of warts, in photography, in the laboratory as an important reagent, etc.

Silver (I) Oxide : Ag_2O; Argentous oxide; A brown solid made by the action of sodium or potassium hydroxide solution to silver nitrate solution. It decomposes on heating to make silver. It dissolves in concentrated ammonia solution to form a complex, $[Ag\ (NH_3)_2]^+$. In its moist state, it is used to hydrolyse alkyl halides.

Silver (II) Oxide ; AgO; Argentic oxide; A black solid made by the oxidation of silver with ozone or by adding a solution of potassium persulphate to silver (I) nitrate.

Sima : The rocks rich in silica and magnesium (the name is derived from these two). Sima is more dense and plastic than sial.

Single Bond : *See chemical bond.*

Sintered Glass : Porous glass used for filtration of precipitates in the gravimetric analysis of certain metals.

S.I. Units : (System International d'Unites) A universally applicable system of units. It has seven basic units – metre for length, kilogram for weight; second for time; kelvin for temperature; ampere for electric current; mole for the amount of substance; candela for luminosity. Derived units may be obtained by multiplication or division of base units.

Slag : A fusible mass formed during the extraction of metals when the impurities of the metal react with the flux added. Slag is generally lighter and floats on the surface of the molten metal, e.g., in a blast furnace the flux used is limestone which removes the main impurity, silica, present in iron ore as calcium silicate slag.

Staked Lime : *See calcium hydroxide.*

Slurry : A thin paste of insoluble substance, e.g., a slurry of silica gel with binder is made in an organic solvent to apply on glass plates for thin layer chromatography

Smelting : An industrial method of extracting metals from the ores at high temperature, using a reducing agent like coke. Carbon itself may reduce zinc and tin oxides while carbon monoxide is used to reduce iron oxide to iron.

Smoke : A colloidal solution of solid carbon particles in air or a gas.

Soaps : Sodium and potassium salts of higher carboxylic acids used as cleansing agents. Soaps may be made by the alkaline hydrolysis (saponification) of vegetables oils and animal fats. Higher carboxylic acids involved may be stearic acid, palmitic acid etc. Potassium soaps are much softer than sodium soaps.

Soaps do not work properly in hard water as the calcium and magnesium salts of hard water form a precipitate with soap. In general soaps are biodegradable.

SODA ASH : *See sodium carbonate.*

SODA LIME : A mixture of sodium hydroxide and calcium oxide. It is used in the laboratory as a drying agent or as a decarboxylative agent. Example, in the preparation of alkanes from carboxylic acids.

SODAMIDE : $NaNH_2$; A white ionic solid made by reacting ammonia with molten sodium. With water it produces ammonia, while on heating with carbon, it forms sodium azanide, and with nitrogen (I) oxide it, forms sodium azide.

SODIUM : Na; A soft grey, highly reactive alkali metal of group IA and 3rd period of the periodic table. Its atomic number is 11 and relative atomic mass 22.98. It occurs widely as sodium chloride in sea water and as deposits of halites in dried up lakes. Other minerals containing sodium are chile saltpetre, Glauber's salt, washerman's clay etc. It is obtained by the electrolysis of molten sodium chloride (Down cell). Sodium is highly reactive and reacts with almost all non metals. Almost all the salts of sodium are water soluble. Sodium hydroxide is a very strong base. Sodium metal is dissolved in liquid ammonia and the solution thus obtained is blue in colour and a good conductor of electricity.

SODIUM ACETATE : *See sodium ethanoate.*

SODIUM ALUMINATE : $NaAlO_2$; A white solid obtained by adding caustic soda solution to alumina, aluminium salt or the metal itself. In solution it exists as $[AL\ (OH)_4]^-$.

SODIUM AZIDE : NaN_3; A while solid made by the action of nitrogen (I) oxide with sodamide. On heating it emits nitrogen.

SODIUM BENZENE CARBOXYLATE : C_6H_5COONa; Sodium benzoate; A white crystalline solid made by neutralising benzoic acid with caustic soda, or sodium carbonate. It is used as a food preservative.

SODIUM BENZOATE : *See sodium benzene carboxylate.*

SODIUM BICARBONATE : *See sodium hydrogen carbonate.*

SODIUM BISULPHATE : *See sodium hydrogen sulphate.*

SODIUM BROMIDE : NaBr; A white crystalline solid prepared by the action of hydrobromic acid on sodium carbonate or by the action of bromine with sodium hydroxide solution. It is soluble in water and is used in medicines and in photography.

SODIUM CARBONATE : Na_2CO_3; Soda ash; A white amorphous solid, soluble in water. It is manufactured from sodium chloride solution by Solvay ammonia process. It exists as $Na_2CO_3 . 10H_2O$ (washing soda). These particles show efflorescene and form $Na_2CO_3 . H_2O$. It is used as a cleanser, as a starting material for the manufacture of caustic soda and as a laboratory reagent.

SODIUM CHLORATE (V) : $NaClO_3$; A white solid made by passing chloride through hot and concentrated solution of sodium hydroxide. It is a strong oxidising agent. On heating, it decomposes to give oxygen. It is used in matches, in explosives, in the textile industry and as an oxidising agent.

SODIUM CHLORIDE : NaCl; Common salt; A white crystalline solid obtained from sea water. It occurs as a rock salt also. It is soluble in water. It is an essential constituent of food as well as preservative. It is used as the starting material for the manufacture of caustic soda, sodium carbonate, soap, etc. Its structure consists of an arrangement of Na^+ ions while ions are situated at the middle of each edge and at the centre of the cube.

SODIUM CYANIDE : NaCN; A white solid formed by the action of carbon sodamide or by the action of ammonia and sodium to form sodamide, which is then reacted with carbon. It is extremely poisonous and is used in the extraction of silver and gold.

SODIUM DICHROMATE : $Na_2Cr_2O_7$; An orange coloured crystalline solid, deliquescent, highly soluble in water. It is manufactured from chromite ore. On heating it decomposes to make chromate, chromic oxide and oxygen. On the addition of alkali, it changes into chromate. It is used as an oxidising agent in volumetric analysis and in organic chemistry.

Sodium Dihydrogen Phosphate (V) : NaH_2PO_4; A white solid prepared by reacting phosphoric acid with sodium hydroxide solution using methyl orange as an indicator. It is used in baking powders.

Sodium Dioxide : *See sodium superoxide.*

Sodium Ethanoate : CH_3COONa; Sodium acetate; A white solid prepared by neutralising acetic acid with sodium carbonate or sodium hydroxide. It is used in the dyeing industry.

Sodium Fluoride : Na; A white crystalline solid, highly soluble in water. Reacts with concentrated sulphuric acid to make hydrogen fluoride. It is used as a constituent of ceramic enamel (mixed with sodium chloride) and as an antiseptic.

Sodium Formate : *See sodium methanoate.*

Sodium Hexaflouroaluminate : Na_3AlF_6; Cryolite; A white or slighty coloured solid substance used as a flux in the manufacture of aluminium metal.

Sodium Hydride : NaH. A white crystalline solid made by passing a current of dry hydrogen through molten sodium in an inert medium. On reacting with water, it forms sodium hydroxide and hydrogen. It acts as a strong reducing agent and may convert sulphuric acid to hydrogen sulphide, iron (III) oxide to iron.

Sodium Hydrogen Carbonate : $NaHCO_3$; Sodium bicarbonate; Baking soda; A white solid manufactured by Solvay process. It decomposes to sodium carbonate on heating. It is used as a constituent of baking powder (mixed with sodium tartarate), in fire extinguishers, etc.

Sodium Hydrogen Sulphate : $NaHSO_4$; Sodium bisulphate; It is prepared by heating sodium chloride and sulphuric acid in equimolar amounts. It is strongly acidic. On strong heating sodium hydrogen sulphate forms pyrosulphate ($Na_2S_2O_7$). It is used in the dyeing industry.

Sodium Hydrogen Sulphite : $NaHSO_3$; Sodium bisulphite; A white solid prepared by passing sulphur dioxide in excess through a solution of sodium carbonate. On heating, it decomposes to give sodium sulphate, sulphur dioxide and sulphur. It is used in medicines as an antiseptic.

Sodium Hydroxide : NaOH; Caustic soda; A white deliquescent solid, soluble in water with release of heat. Its aqueous solution is strongly alkaline, soapy in touch and highly corrosive. It is manufactured by the electrolysis of sodium chloride solution (Castner Kellner cell and Nelson cell). It is used in the manufacture of soap and paper. In the laboratory and industry it is used to absorb acidic gases like carbon dioxide, sulphur dioxide.

Sodium Iodide : NaI; A white solid made by the action of hydroiodic acid on sodium carbonate or sodium hydroxide. It is used in medicine and photography.

Sodium Methanoate : HCOONa; Sodium formate; A white solid made by reacting carbon monoxide and solid sodium hydroxide under high pressure and at high temperature.

Sodium Monoxide : Na_2O; A white solid obtained by burning sodium in a small amount of air or oxygen. It reacts violently with water to form caustic soda solution. With liquid ammonia it forms sodamide and sodium hydroxide.

Sodium Nitrate : NaN_4O_3; Chile saltpetre; A white solid formed by the neutralisation of nitric acid with sodium carbonate or sodium hydroxide. It occurs as such in the earth's crust. On heating, it decomposes to produce oxygen and sodium nitrite. It is a source of nitric acid in the laboratory. It is used as a fertiliser.

Sodium Nitrite : $NaNO_2$; A yellowish white solid obtained by heating sodium nitrate. It is readily soluble in water. It is used in the laboratory for diazotisation. In industry, it is used as a corrosion inhibitor.

Sodium Orthophosphate : *See trisodium phosphate (V).*

Sodium Peroxide : Na_2O_2; A yellowish white solid made by burning sodium in excess of oxygen. It is used as a bleaching agent for wool and wood pulp. It is used to make hydrogen peroxide in the laboratory.

Sodium Sulphate : $NaSO_4$; A white solid which exists in nature as its decahydrate, Glauber's salt ($NaSO_4.10H_2O$). Sodium sulphate decahydrate is used in the manufacture of glass and as a purgative in medicine. In air, it loses water and forms a anhydrous salt (efflorescence).

SODIUM SULPHIDE : Na_2S; A reddish solid formed by reducing sodium sulphate, using carbon or carbon monoxide. It is deliquescent in nature and highly soluble in water. The solution is strongly alkaline and acts as solution of H_2S in qualitative analysis.

SODIUM SULPHITE : Na_2SO_3; A white solid formed by the action of sodium carbonate or sodium hydroxide with sulphur dioxide in equal amounts. It crystallises as heptahydrate.On reacting with dilute mineral acids, it emits sulphur dioxide.

SODIUM SUPEROXIDE ; NaO_2; A yellowish solid prepared by heating sodium peroxide in oxygen. With water it forms hydrogen peroxide, sodium hydroxide and oxygen.

SODIUM THIOSULPHATE : $Na_2S_2O_3$; A white solid crystalline compound prepared by boiling sodium sulphite with sulphur. It crystallises as its pentahydrate. With dilute mineral acids, it forms sulphur and sulphur dioxide. It is used as a fixing agent in photography and as antichlor in the textile industry. It is used in iodometric titrations.

SOFT SOAP : A soap made by saponification of oil and fats with potassium hydroxide.

SOFT WATER : Water having no soluble salts of calcium and magnesium.

SOL : A colloidal solution of solid dispersed phase and liquid dispersion medium, e.g., metal sols, ferric hydroxide sol, arsenous sulphide sol, etc. They are of two types – lyophilic and lyophobic sols (discussed separately).

SOLDER : An alloy of tin and lead used in joining metals. Soft solder has nearly 60% tin and is used for making electric contacts. Brazing solders, copper-zinc alloys, melt at a higher temperature but make strong joints.

SOLID : The state of matter in which the constitutent particles, atoms, molecules or ions, are very closely held, giving it a definite shape and size. They may be crystalline or amorophous. Crystalline solids may have different crystal lattices. Properties of solids are associated with the type of bonds present and the type of lattice.

SOLID SOLUTION : A solid composed of two of more substances mixed together at the molecular level. Constitutents of one solid may occupy the lattice positions occupied by the other. Certain alloys are solid solutions of one metal in another. Example, Ni-Cu alloys.

SOLUBILITY : Maximum amount of a substance which may be dissolved in the given amount of a solvent (to form a saturated solution). Its value is fixed at a particular temperature and pressure. Units may be moles of solute per 100 g of solvent or number of grams per litre of solution.

SOLUBILITY PRODUCT : K_{sp}; It is the product of active masses of ions formed in a saturated solution, raised to suitable powers equal to the number of each type of ion formed per molecule of electrolyte. For calcium fluoride, CaF_2, it is given by $K_{sp} = [Ca^{++}]\ [F_2^-]^{2}$. Values of solubility product help in calculating solubility of a substance and in predicting the precipitation. If ionic product exceeds the solubility product the substance gets precipitated.

SOLUTE : A material that is dissolved in a solvent to form a solution. If the solute and solvent are in the same phase, the substance in smaller proportions is called solute.

SOLUTION : A homogenous mixture of two or more substances in the same or different phases. The component in larger proportions is called solvent.

SOLVATION : Direct intermolecular interaction of the solvent with the ions or molecules of the solute. The energy released during solvation (hydration energy in case of water) is the force which overcomes the attraction between the positive and negative ions and dissolves the solid.

Solvay Process : A process for the manufacture of sodium carbonate from sodium chloride solution with the help of limestone and a little ammonia (to initiate the process). Limestone is the source of carbon dioxide and lime.

Solvent : A liquid capable of dissolving other substances to form a solution. Solvent may be polar or non polar. A polar solvent (water) can easily dissolve ionic substances while a non polar solvent (benzene etc.) can dissolve covalent, organic substances. Another classification of these solvents may be protic solvent (H_2O, HF, NH_3 etc.) and aprotic solvents (CCl_4, CS_2 etc.) They may also be divided into aqueous (H_2O) and non aqueous solvents (ammonica, sulphur dioxide, hydrogen fluoride in liquid state).

Solvent Extraction : A process of extracting a substance from a mixture by shaking it with a particular suitable solvent.

Solvolysis : Interaction between a solvent and a given substance dissolved (hydrolysis, when the solvent is water).

Sorption : Absorption.

Specific : Describing a physical quantity per unit mass, e.g., specific volume, specific heat, etc.

Specific Gravity : *See relative density*

Spectral Line : A radiation of a particular wavelength emitted or absorbed by an atom, ion or molecule. *See line spectrum also.*

Spectral Series : A group of related lines in the absorption or emission spectrum, e.g., spectral lines of various series in the hydrogen spectrum.

Spectrograph : A device producing a photographic record of a spectrum.

Spectrometer : An instrument for the investigation of various electromagnetic radiations. A spectrometer has a source of radiation, a prism or grating, a photocell. Also called spectro scopes.

Spectrophotometer : A type of spectrometer able to measure the intensity of radiation at different frequencies in the given spectrum

SPECTROSCOPE : *See spectrometer.*

SPECTROSCOPY : The production and investigation of spectra. It is used for the analysis of mixtures, identification of compounds and for determining the structure of chemical compounds.

SPECTRUM : An arrangement of electromagnetic radiations emitted or absorbed by a susbstance. In an emission spectrum, light radiations are emitted by a substance due to high temperature, bombardment of electrons, absorption of higher frequency radiation etc. In an absorption spectrum, a continuous flow of radiation is passed through the sample and the radiations are then analysed to determine the absorbed wavelength.

SPECULUM : An alloy of copper and tin previously used in the reflecting telescope to make the main mirror (reflecting surface).

SPELTAR : Commercial variety of zinc having about 3% of lead impurity.

SPHALERITE : ZnS; Zinc blende; A mineral form of zinc sulphide.

SPIEGEL; SPIEGELEISEN : A form of pig iron containing about 15-30% manganese and 4-5% carbon. It is added to steel to raise the amount of manganese in steel.

SPIN : A property of electron like a particle revolving round its own axis, so that the particle has an angular momentum. The values of spin quantum number for an electron may be + and –.

SPINEL : A class of mixed metal oxide with the general formula $A^{2+} B_2{}^{3+} O_4$ where A^{2+} may be Mg^{++}, Fe^{++}, Zn^{++}, Mn^{++} or $N_1{}^{++}$ while B^{3+} may be Al^{3+}, Fe^{3+} or Cr^{3+}

SPIRIT OF SALT : Hydrochloric acid; It is made by reacting sodium chloride with concentrated sulphuric acid.

SPONTANEOUS COMBUSTION : Burning of a compound without requiring any heat from an external source. By a slow oxidation process, sufficient heat is produced within itself.

SPONTANEOUS REACTION : A reaction which proceeds on its own without any external assistance. Spontaniety of the reactions is governed by Gibbs-Helmoltz equation,

$\Delta G = \Delta H - T\Delta S$ where, ΔH is the enthalpy change and ΔS is the entropy change. Reaction is which $\Delta G < O$ are spontaneous. If $\Delta G = O$ the reaction is said to be in the equilibrium state.

Square Planar : Describing the geometry in which the central atom is surrounded by four groups having 90° angles with one another and directed towards four corners of a square.

Stabilisation Energy : The difference in energy between the delocalised structure and the conventional structure. It may also be described as resonance energy. For example, the stabilisation energy of benzene is 150 K J $mole^{-1}$.

Staggered Conformation : *See conformation.*

Stainless Steel : *See steel.*

Standard Cell : A galvanic cell whose e.m.fs is used as standard, e.g., Weston cadmium cell.

Standard Electrode : A half cell or electrode whose electrode potential value is standard and may be used to determine the electrode potentials of other half cells, e.g., normal hydrogen electrode and calomel electrode.

Standard Pressure And Temperature : An internationally accepted value, of pressure 760 mm Hg and temperature 273 K.

1 atmosphere pressure = 76 cm Hg = 101325 Pa.

273 K temperature = $O^{o}C$.

Standard Solution : A solution whose strength is known, i.e., a solution made by dissolving a known weight of solute. Consequently, normality and molarity of such solutions are known and do not change with time. If the substance is not available in pure state or is hygroscopic in nature, the solution made must be standardised by titrating against another standard solution.

Standard State : Describing the standard conditions used in thermodynamics. By standard conditions we mean pressure 101325 Pa (1 atmosphere), temperature 298 K and concentration of 1mole.

STANDARD TEMPERATURE : An internationally agreed value for which several measurements are described is 273 K.

STANNANE : SnH_4; Tin (IV) hydride; A colourless gaseous substance prepared by treating $LiAlH_4$ on $SnCl_2$. It is unstable, poisonous and reducing in nature.

STANNIC COMPOUNDS : Compounds of tin (IV).

STANNOUS COMPOUNDS : Compounds of tin (II)

STARCH : A polysaccharide which occurs in maize, wheat, barley, potato etc. By enzyme action, they may be decomposed to simple sugars and then give energy by metabolism. It is made of two components amylose (water soluble) and amylopectin (water insoluble). On complete hydrolysis, starch decomposes to glucose.

STATES OF MATTER : The physical states, solid, liquid and gas, in which a substance may exist. Plasma is considered as the fourth state of matter.

STEAM DISTILLATION : A method of purification and isolation of a substance in which steam is passed through impure compounds or mixtures. Vapours of a compound are distilled out along with steam and the distillate is a mixture of water and the compound. Such a distillation takes place below 373K. The compound to be distilled by this method should be immisicible with water.

STEARATE : A salt or ester of stearic acid (octadecanoic acid).

STEARIC ACID : *See octadecanoic acid.*

STEEL : An alloy of iron with varying amounts of carbon (upto 1.5%) along with small amounts of other elements (alloy steels) like silicon, chromium, manganese, nickel, molybdenum, etc. Carbon steels exist in three stable crystalline forms—ferrite (bcc) austenite (fcc) & cementite (orthorhombic crystalline). Pearlite is a mixture of ferrite and cementite arranged in parallel plates. Steel may be manufacture by Bessemer process and open hearth process, but the latest is the L.D. process. Properties of the steel depend on its composition, e.g., stainless steel (chromium upto 25%) is corrosion resistant.

Step : An elementary stage in some complex reaction. A complex reaction may involve several steps of which the slowest step is the rate determining step.

Steradian : Sr; The S.I. unit of solid angle.

Stereochemisty : The branch of chemistry dealing with the study of arrangement of groups or atoms in space and the manner in which the properties depend on such an arrangement.

Stereoisomerism : Isomerism concerned with the different arrangements of atoms and groups in space, e.g., geometrical and optical isomerisms.

Steric Effect : Due to the size of groups, shapes and properties of molecules are influenced. The approach of a reagent may be hindered due to the presence of bulky groups.

Steric Hindrance : *See steric effect.*

Steroids : A term describing the substances having carbon skeleton of sterols, e.g., bile acids, sex hormones, etc.

Still : An apparatus used for distillation.

Stoichiometry : The ratio in which the elements form a compound. It is generally a small whole number ratio.

Storage Battery : *See accumulator.*

Stp (Ntp) : Standard temperature and pressure. The values are 101325 Pascals pressure and 273 K (exactly 273. 15K) temperature.

Straight Chain : *See chain*

Strecker Synthesis : A method of preparing aminoacids through cyanohdyrins.

Strotia : *See strontium oxide.*

Strontium : Sr; A soft, reactive, low melting alkaline earth metal of group IIA and 5th period. Its atomic number is 38 and relative atomic mass is 87.62. It occurs in the earth's crust as strontianite ($SrCO_3$) and celestine ($SrSO_4$). It forms a variety of compounds by directly combining with oxygen, nitrogen, halogen, sulphur and hydrogen. Its oxide and hydroxide are baisc in nature. Its carbonate and sulphate are insoluble like those of barium. It was discovered by Klaproth and Hope.

Strontium Bicarbonate : *See strontium hydrogen carbonate.*

Strontium Carbonate : $SrCO_3$; A white insoluble substance, prepared by passing carbon dioxide through SrO, $Sr(OH)_2$ or solution of its salt. It is used in fireworks to produce red colour in the flame.

Strontium Chloride : $SrCl_2$; A white solid prepared by the action of chlorine and the metal or metal oxide.

Strontium Hydrogen Carbonate : $Sr(HCO_3)_2$; Strontium bicarbonate; A white solid prepared in solutions by passing carbon dioxide through the water suspension of strontium carbonate. On heating it forms strontium carbonate.

Strontium Hydroxide . $Sr(OH)_2$; A white solid prepared by dissolving strontium oxide in water and crystallising it in octahydrate ($Sr(OH)_2$. $8H_2O$) Its solution is strongly basic.

Strontium Oxide : SrO; Strontia; A greyish white solid obtained by heating $Sr(OH)_2$, $SrCO_3$ or $Sr(NO_3)_2$. It dissolves in water to make hydroxide.

Strontium Sulphate : $SrSO_4$; A white solid, very little soluble in water, occurring as celestine in nature. It is prepared by the action of sulphuric acid on oxide, hydroxide or carbonate of the metal.

Structural Formula : The exact formula of a compound representing the number of types of atoms present, types of bonds involved and the arrangement of atoms as far as possible.

Structural Isomerism : *See isomerism.*

Styrene : *See phenylethene.*

Sublimate : A solid obtained as a result of sublimation.

Sublimation : The change of a solid substance directly into its vapour phase without changing into its liquid phase below the melting point of the solid, e.g., iodine, camphor etc.

Subshell : A part of the shell which describes the shape of the orbitals concerned. It is represented by the values of angular momentum quantum number. A shell has a certain definite number of subshells, e.g., the 3rd orbit has three subshells namely 3s, 3p and 3d. Orbitals belonging to a particular subsnell are always degenerate, e.g., five orbitals of 3rd subshell have equal energies.

SUBSTITUENT : An atom or group of atoms substituted for another in a compound, e.g., halogen groups, nitro groups are substituents which may be substituted in place of hydrogen.

SUBSTITUTION REACTIONS : The reactions showing the replacement of one or more atoms in a compound by other atoms or groups. They may be of various types – electrophilic (in benzene), nucleophilic (alkyl and aryl in alkanes).

SUBSTRATE : A substance which reacts with a catalyst or enzyme.

SUCCINIC ACID : *See butanedioic acid.*

SUCROSE : $C_{12}H_{22}O_{11}$; Cane sugar; A disaccharide which occurs in plants and, on hydrolysis, it decomposes to glucose and fructose. Hydrolysis takes place in the presence of the enzyme invertase and is called inversion. Sucrose is dextrorotatory. It is manufactured from sugarcane and beets.

SUGAR : A class of carbohydrates. Sugars are water soluble and sweet in taste. They are polyhydric compounds having an aldehyde or a ketone group. Also known as saccharides, they may be classified as monosaccharide (glucose, fructose, ribose etc), disaccharides (sucrose, maltose etc). Based on the presence of a functional group they may be classified as aldoses and ketoses. Structurally, they are shown by an open chain structure as well as a ring structure, e.g., glucose exists as a six membered heterocyclic ring (pyranose ring), while fructose exists as a five membered ring (furanose ring).

SULPHA DRUGS : *See sulphonamide.*

SULPHATE : A salt or ester of sulphuric acid.

SULPHIDE : S; A substance made by the action of hydrogen sulphide on the suitable salt solution. They contain 5^- ions or groups.

SULPHITE : SO_3^-; A salt or ester of sulphurous acid.

SULPHONAMIDE : A group of compounds having the general formula RSO_2NH_2, used in making several sulpha drugs.

SULPHONATION : A reaction by which $-SO_3H$ group is introduced in an organic compound, generally benzene. Reaction takes place by heating the compound, with concentrated sulphuric acid. The attacking electrophile for sulphonation is SO_3.

SULPHONIC ACID : A class of organic compounds containing -SO_2-OH group. They are made by sulphonation, e.g., $C_6H_5SO_2$. OH, benzene sulphonic acid. The group when present on the benzene ring acts as a deactivating group and directs the incoming group to meta position.

SULPHUR : S; A yellow coloured non metal of the 16th group and 3rd period of the periodic table. It has atomic number 16 and relative atomic mass 32.06. Occurs as sulphide and sulphate ores. Exists in various allotropic forms like alpha, beta, gamma sulphur. It shows various oxidation states in its compounds ranging from -2 to +6. In the family it shows maximum catenation.

SULPHUR DICHLORIDE DIOXIDE : SO_2Cl_2; Sulphuryl chloride; A colourless, fuming liquid formed by the reaction between chlorine and sulphur (IV) oxide.

SULPHUR DICHLORIDE OXIDE : $SOCl_2$; Thionyl chloride; A colourless fuming liquid obtained by reacting sulphur (IV) oxide and phosphorus (V) chloride. It is used as a chlorinating agentm, e.g., alcohols are changed into alkyl halides by this reaction.

SULPHURETTED HYDROGEN : *See hydrogen sulphide.*

SULPHURIC (IV) ACID : *See sulphurous acid.*

SULPHURIC (VI) ACID : H_2SO_4; Oil of vitriol; A colourless, oily liquid made by the contact process using Pt or V_2O_5 catalyst. The concentrated acid is a very powerful oxidising agent and a strong deyhdrating agent. It is a dibasic acid and may form two series of salts called bisulphate (hydrogen sulphates) and sulphates. It is mainly used to make fertilisers, other chemicals, paints, detergents and fibres. Due to its very high reactivity it is also known as the king of chemicals.

SULPHUR MONOCHLORIDE : S_2Cl_2; *See disulphur dichloride.*

SULPHUROUS ACID : H_2SO_3; Sulphur (IV) acid; A weak dibasic acid found only in solution. Prepared by dissolving sulphur (IV) oxide in water. It makes two types of salts sulphites and hydrogen sulphites. It acts as a reducing agent and may reduce iron (III) salts, chlorine, dichromate ions etc.

Sulphur (VI) Oxide : SO_3; Sulphur trioxide. A white fuming volatile solid prepared by the oxidation of sulphur (IV) oxide in the presence of catalyst vanadium (V) oxide. It is an important intermediate of sulphuric acid. With water it reacts violently to make sulphuric acid. It dissolves in sulphuric acid to make oleum or fuming sulphuric acid.

Sulphur Troxide : *See sulphur (VI) oxide.*

Sulphuryl Chloride : *See sulphur dichloride dioxide.*

Superconduction : Property of zero resistance exhibited by certain metals and alloys at very low temperatures. Attempts are being made to produce super conductors at room temperature to use them in power transmission, aviation, transportation etc.

Super fluidity : A property of helium at very low temperature. It is the property associated with high thermal conductivity and movement without friction.

Super heating : The rise of the temperature of a liquid above its boiling point. It is achieved at a pressure more than the atmospheric pressure.

Super Oxide : An inorganic compound having O_2^- ions.

Super Plasticity : The ability of some metals and alloys to stretch extremely at temperature, unlike normal alloys. It was first observed in an alloy of zinc and aluminium.

Super phosphate : A mixture of calcium sulphate and calcium hydrogen phosphate. It is made by the action of sulphuric acid on calcium phosphate. It is used as a fertiliser.

Supersaturated Solution : A saturated solution of a solid which possesses more solute for some time. It is made by cooling the saturated solution.

Supplementary Units : The dimensionless units like radian and steradian used alongwith fundamental units to make derived units.

Surfacant : A substance having the property of wetting and lowering surface tension. Surface action agents, as they are called, may be soaps, detergents etc.

SUSPENSION : A heterogenous system of a solid in liquid, with the particles of the solid dispersed in a liquid. They generally settle down on keeping.

SYLVINE : A mineral form of KCl.

SYNTHESIS : Preparation of a compound from its simpler constituents.

SYNTHETIC : Describing a compound made artificially.

SYSTEM INTERNATIONAL d' UNITS : See S.I. units.

T

TALC : $Mg_3(OH)_2\ Si_4O_{10}$; French chalk; A silicate used as a filler in paints, rubber, insecticides and as a constituent of toilet and cosmetic powder.

TANNIC ACID : A yellowish complex organic compound present in several plants. It is used in dyeing as a mordant.

TANNIN : One of a class of organic chemicals found in leaves, unripe fruits and the bark of tree. Some are used in tanning, in the production of ink and as a mordant in the textile industry.

TANTALUM : Ta; A blue grey heavy transition metal belonging to the 5th group and 6th period of the periodical table. Its atomic number is 73, while relative atomic mass is 180.95. It is found with niobium in the ore columbite-tantalite (Fe,Mn) (Ta, Nb)O_{26}. Due to its high resistivity against corrosion it is used in turbine blades, surgical instruments and pins to join bones. It was first isolated by Berzelius.

TAR : Any of the various black semi–solid mixtures of hydrocarbons along with free carbon, obtained as a residue in the destructive distillation of coal or petroleum.

TARTARIC ACID : $HOOC(CHOH)_2COOH$; A white crystalline, dibasic, dicarboxylic organic acid. It is optically active and may exist in the form of two enantiomers and one optically inactive meso form. Its IUPAC name is 2,3-dihydroxybutanedioic acid. It is used in baking powder and other foodstuff.

TARTRATE : An ester or salt of tartaric acid.

TAUTOMERISM : A type of isomerism in which two isomeric forms of a compound can convert into each other to make an equilibrium mixture. It results from the migration of a hydrogen atom and pi-electron density, e.g., keto–enol tautomerism in aldehydes and ketones.

TECHNETIUM : Tc; A radioactive transition metal of group VIIB and 5th period of the long form of periodic table. Its atomic number is 43 and relative atomic mass is 97. It can be obtained artificially by bombarding molybdenum with neutrons. It is also present in the fission products of uranium. It was made for the first time by Perrier and Segre.

TEFLON : Trade name for polytetrafluoroethane (PTFE).

TELLURIDES : Binary substances made of tellurium with some other electropositive elements, e.g., H_2Te.

TEMPERATURE SCALES : Several practical scales of temperature are in use of which the centigrade or Celsius scale is most common. In some countries, Fahrenheit scale is still used. The scale is developed by fixed points as the freezing point of pure water and its boiling point, while the gap is divided into a definite number fractions (100 in case of Celsius and 180 in case of Fahrenheit). In the latest International Practical Temperature Scale there are 11 fixed points that cover the range from 13.81 K to 1337.58K.

TEMPERING : The process of increasing the toughness of steel by heating to a definite high temperature and cooling it slowly to room temperature. The rate of cooling is maintained away to give the desirable properties to steel.

TEMPORARY HARDNESS : *See hardness of water.*

TERA : T; A prefix denoting 10^{12}. Example, $1\ Tm = 10^{12}m$

TERBIUM : Tb; A soft, ductile, silvery metal belonging to the lanthanide series (4f series) of elements. Its atomic number is 65 and relative atomic mass is 158.93. It occurs in apatite and xenotine alongwith other rare earths. it was discovered by C.G. Mosander.

TERNARY COMPOUND : A chemical compound made by the combination of three elements, e.g., sodium sulphate, potassium chlorate etc.

TERPENES : A class of unsaturated hydrocarbons present in plants. they generally contain isoprene unit $CH_2 = C(CH_3)\ CH = CH_2$, with the general formula $(C_5\ H_8)n$.

TERTIARY ALCOHOLS : Alcohols having general formulae R_3COH.

TERTIARY AMINES : Amines having general formulae R_3N.

TERVALENT : Having a valency of three (trivalent)

TERYLENE : Trade name for polyesters.

TESLA : T; The S.I. unit of magnetic flux density, equal to the flux density of one weber per square metre. $1T = Wbm^{-2}$

TETRACHLOROMETHANE : CCl_4; Carbon tetrachloride; *See carbon tetrachloride for details.*

TETRAETHYL LEAD : *See lead tetraethyl.*

TETRAHEDRAL ANGLE : The angle between the bonds in a tetrahedral compound. It is 109.280 in purely tetrahedral substances.

TETRAHEDRAL COMPOUND : A compound in which the central atom has four valencies directed towards the corners of a regular teterahedron.

TETRAHYDRATE : A crystalline compound having four molecules of water of crystallisation per molecule.

TETRAHYDROFURAN : THF; A colourless liquid used as a solvent. It has a heterocyclic ring structure (C_4H_8O)

THALLIUM : Tl; A soft greyish metal of group IIIA and 6th period of the periodic table. Its atomic number is 81 and relative atomic mass is 204.37. It is found in lead and cadmium ores like pyrites (FeS_2). Its sulphate was used as a rodenticide. Thallium (III) compounds are easily reduced to Thallium (I) state and so act as strong oxidising agents. Due to inert pair effect, the most common state is +1. It was discovered by William Crookes.

THERMITE : A mixture of iron (III) oxide and aluminium powder used for the welding of heavy structures. The reaction is highly exothermic and produces iron in the molten state.

THERMOCHEMISTRY : The branch of chemistry dealing with the heats of reactions of various types.

THERMODYNAMICS : The branch of science dealing with the laws governing the interconversion of energy from one form to the other, the direction of heat flow and the capability of a system to do work. It may briefly be described in the form of certain laws :

First law of thermodynamics states that total energy of an isolated system remains constant, although one form may change into another form.

Second law deals with the feasibility of processes and states that heat cannot transfer itself from a cold to a hot body.

The third law of thermodynamics states that for a perfect crystalline substance at absolute zero the entropy is zero.

Zeroth law states that if two bodies are each in thermal equilibrium with a third body, then all the three bodies are in thermal equilibrium with each other.

THERMODYNAMIC TEMPERATURE : T; A temperature measure in Kelvin.

THERMOLUMINESCENCE : Luminescence produced in a solid when its temperature is raised.

THERMOMETER : A device used for measuring the temperature of a system. The most simple device is a liquid in a glass tube, a thermometer based on the expansion of liquid. It contains mercury or alcohol. Another device, a bimetallic thermometer consists of two metal strips bonded together and coiled. It is based on the unequal expansion of two metals. The Beckmann thermometer is used to measure the small variations.

THERMOPLASTIC POLYMER : *See plastics.*

THERMOSETTING POLYMER : *See plastics.*

THIAMINE : *See vitamin B–complex.*

THIN LAYER CHROMATOGRAPHY : TLC; Technique used for the analysis of liquid mixtures. It employs a stationary phase as alumina or silica gel spread evenly on a glass slide. A small amount of mixture is employed on the lower part of the glass plate with the help of a capillary table. It is now kept straight in the solvent which rises upward by capillary action. Due to the difference in the tendency of the components to move, separate layers may be formed. After the electron plate is dried and developed, either by spraying with a suitable chemical or by UV light. The components are identified by comparing R_f values with those of the standard compounds.

THIOALCOHOLS : RSH; Mercaptans; Compounds in which the O atom of the alcohol is replaced by sulphur. Ethyl mercaptan is mixed with LPG in small amounts to detect leakage (due to its peculiar smell).

THIOETHERS : RSR; Compounds in which the O atom of ether replaced by sulphur.

THIOSULPHATE : $S_2O_3^-$ ion; A salt containing $S_2O_3^-$ ion.

THIONYL CHLORIDE : *See sulphur dichloride oxide.*

THIXOTROPY : The change of viscosity of a liquid on standing over a period of time. Liquids showing this property are called thixotropic, e.g., paints.

THORIUM : Th; A poisonous, silvery, ductile, radioactive element of the actinide series (5f series). It has atomic number 90 and relative atomic mass 232.04. Its most important mineral is monazite. It is used in magnesium alloys and in the electronic industry. It can be used as a nuclear fuel for breeder reactors. Thorium is used in refractories also. It was discovered by Berzelius.

THULIUIM : Tm; A soft grey metal of the lanthanide series, having atomic number 69 and relative atomic mass 168.934. It occurs in apatite and xenotine along with other lanthanides. It was discovered by P.T. Cleve.

THYMINE : A pyrimidine base and a major constituent of DNA.

TIN : Sn; A white, shining metal of low melting point belonging to the 14th group and 5th period. Its atomic number is 50 and relative atomic mass is 118.7. It is found as cassiterite (SnO_2) and is extracted by reduction with carbon. It is a constituent of several alloys like phosphous bronze, gun metal, solder and pewter. It is used for tin plating. Tin is a reactive metal which combines with chlorine and oxygen directly. It is soluble in acids as well as alkalies. It exhibits +2 and +4 oxidation states. Tin has three crystalline allotropes, alpha tin or grey tin, beta tin or white tin and gamma tin

Tin (II) Chloride : $SnCl_2$; A transparent, lustrous solid made by dissolving tin in hydrochloric acid. It is used as a mordant and a reducing agent in acid solutions. May be used for reducing Fe^{3+} and Hg^{2+} compounds.

Tin (IV) Chloride : $SnCl_4$; A colourless fuming liquid. It is hydrolysed by water with hydrochloric acid and forms $SnCl_6^{2-}$.

Tin (IV) Hydride : *See stannane.*

Tin (II) Oxide : SnO; A dark coloured solid made by precipitating the hydrated oxide from a solution containing tin (II) ions and heating the product to 373K.

Tin (IV) Oxide : SnO_2; A colourless crystalline solid, insoluble in water. It occurs in nature as cassiterite. It is amphoteric in nature.

Tin (IV) Sulphide : SnS_2; A yellowish solid prepared by passing hydrogen sulphide gas through a solution of tin (IV) salt. Its crystalline form is called mosaic gold, which is formed by heating a mixture of tin fillings, sulphur and ammonium chloride. It is used as a pigment.

Titanium : Ti ; White transition metal of the 4th group and 4th period of the period table. Main ores of titanium are rutile (TiO_2) and ilmenite ($FeTiO_3$). It is used in alloys for aircraft, ships, etc., as its alloys are strong, light and corrosion free. At high temperature, it reacts with oxygen, nitrogen, chlorine and other non-metals. It exhibits +2, +3 and +4 oxidation states. It was discovered by Gregor.

Titanium (IV) Chloride : $TiCl_4$; A volatile colourless liquid formed by heating TiO_2 with carbon in the presence of dry chlorine at 973 K. In moist air it forms oxychlorides. It undergoes hydrolysis in water. Its crystalline form is $TiCl_4.5H_2O$.

Titanium (IV) Oxide : TiO_2; A white solid which occurs in three crystalline forms rutile (tetragonal), brokite (orthorhombic) and anatase (tetragonal). It is amphoteric and is used as a white pigment.

Titrant : *See titration.*

Titration : A volumetric method in which a solution of known strength is added to a solution of unknown concentration with the help of a burette until an end point is reached. End point may be determined by adding a suitable indicator. The solution whose concentration is known is called titrant.

TNT : *See trinitrotoluene.*

TOLLEN'S REAGENT : A complex made by adding excess of ammonia solution to an aqueous solution of silver nitrate. It contains $[Ag(NH_3)_2]^+$ ions. It is used in the silver mirror test for to deduct aldehydes.

TOLUENE : C_6H_5-CH_3; Methyl benzene.

TONNE : A unit of mass. 1 tonne = 10^3 kg.

TORR : A unit of pressure. 1 torr = 1 mm Hg. 1 torr = 101325/760 pascals.

TRACER : An isotope of a particular element used to study the mechanism of a particular reaction.

TRANS : Describing a geometric isomer having similar groups on opposite sides of the references bond.

TRANSACTINIDE ELEMENTS : Elements having atomic numbers 104 and 105, i.e., beyond actinides. All these are unstable, short-lived, highly radioactive and are synthetic.

TRANSITION ELEMENT : A class of elements in the periodic table positioned between the s and p-blocks. They are distributed into three series: 3d series $_1$Sc to $_0$Zn; 4d series $_9$Y to $_9$Cd and 5d series $_7$La to $_0$Hg. Fourth series is incomplete and has only three elements. Characteristic properties of these elements are variable oxidation states, formation of coloured compounds, formation of complexes, paramagnetic nature of compounds, etc.

All of them are typical metals. They form various types of alloys. Many of them act as catalysts for several industrial processes.

TRANSITION STATE : An intermediate short-lived, high energy molecule or radical formed during a reaction between the reactants and products. Transition state has higher energy than those of the reactants and products. In this state the old bonds existing in between the atoms of reactants are in the process of breaking while the new bonds of the atoms between the product molecules are being formed. Such a transition state may either dissociate to form the products or break down to form the reactants back.

TRANSITION TEMPERATURE : A definite constant temperature at which a particular physical change occurs in the given compound. It may be change of state (Like vaporisation, fusion, sublimation etc.) or change from one allotropic form to the other.

TRANSMUTATION : The conversion of one element into another by the bombardment of the nuclei with particles or by radioactive disintegration.

TRANSPORT NUMBER : Transference number. It is the fraction of the total charge carried by the cations or anions in the solution of given electrolyte. The more the transport number, the more is the velocity of migration of that type of ion.

TRANSURANIC ELEMENTS : Elements having atomic numbers greater than 92. These elements in general are highly radioactive, short-lived and do not occur in nature as such. They are obtained by the high energy bombardment of neutrons on the lower actinides like thorium and uranium. Transuranic elements cannot be isolated from one another due to very short lives.

TRIATOMIC : Denoting a molecule, radical or ion made of three atoms, e.g., O_3, CO_2, H_2O.

TRIBROMOMETHANE : $CHBr_3$; Bromoform; A colourless liquid.

TRICHLOROACETIC ACID : *See trichloroethanoic acid.*

TRICHLOROETHANOL : Cl_3CCHO; Chloral; A colourless liquid made by the chlorination of ethanol. It is an intermediate in the preparation of chloroform. It is the starting material for manufacturing DDT. It is used as a sedative.

TRICHLOROETHANEDIOL : $Cl_3C.CH(OH)_2$; Chloral hydrate; It is obtained by the hydrolysis of chloral.

TRICHLOROETHANOIC ACID : CCl_3COOH; Trichloroacetic acid; Made by the chlorination of acetic acid in the presence of red phosphorous.

TRICHLOROMETHANE : $CHCl_3$; Chloroform; A colourless, volatile liquid, formerly used as anaesthetic. It is prepared by the action of bleaching powder and ethanol or bleaching powder and acetone. It is used as a solvent.

TRICLINIC CRYSTAL : *See crystal system.*

TRIGONAL BIPYRAMID : A type of geometry in which three valencies are directed towards the corners of an equilateral triangle while the remaining two are directed up and below the plane of the molecule. It is explained by sp^3 hybridisation. Bond angles are 90° and 120°, for example, PCl_5.

TRIGLYCERIDE : An oil or fat in which all the three -OH groups of glycerol are acylated. It is a triester. It is used to make soaps of various types as well as glycerol.

TRIHYDRATE : A crystalline compound having three molecules of water of crystallisation per molecule.

TRIHYDRIC ALCOHOL : Triol; An alcoholic compound having three -OH groups attached to three different carbon atoms, e.g., glycerol.

TRIODOMETHANE : CHI_3; Iodoform; A yellow crystalline solid made by the action of ethanol (or acetone) with an alkaline solution of iodine. The reaction is given by all methyl ketones and with secondary alcohols having $RCH(OH)CH_3$, besides ethanol. In contact with skin it liberates iodine and so is used as an antiseptic in skin ointments.

TRIIRON TETROXIDE : Fe_3O_4; Ferrosoferric oxide; Magnetic oxide of iron; A black coloured solid prepared by the action of steam on red hot iron. It occurs in the mineral, magnetite. It is insoluble in water, but dissolves in acids.

TRIMER : A molecule formed by the combination of three identical molecules.

TRIMETHYL ALUMINIUM : $(CH_3)_3Al$; A colourless liquid. The compound is highly reactive and burns in air. It reacts vigorously with water, acids, halogens, alcohols, etc. They are used as catalyst in Ziegler process.

TRIMOLECULAR : Denoting a reaction which involves the simultaneous collision of three molecules, atoms or ions. The probability of such a reaction is very low.

TRINITROTOLUENE : $(CH_3)C_6H_2(NO_2)_3$; TNT; A yellow crystalline, highly unstable compound, used as an explosive. It is made by the nitration of toluene.

TRID : *See trihydric alcohol.*

TRIOXYGEN : *See ozone.*

TRIPLE BOND : A covalent bond formed between two atoms, in which three pairs of electrons make a bond (shared between the atoms). It is a combination of one sigma and two pi-bonds, e.g., N=N; HC=CH.

TRIPLE POINT : The point at which the gas, solid and liquid phases of a substance can coexist in equilibrium. For water its value is 273.16 K at 101325 P.

TRISODIUM PHOSPHATE (V) : Na_3PO_4; Sodium orthophosphate; A white solid prepared by adding sodium hydroxide to disodium hydrogen phosphate. It exists in the hydrated crystalline form as $Na_3PO_4.12H_2O$. It is neither deliquescent not efflorescent. It is used as a water softener.

TRITIATED COMPOUND : A compound in which one or more $_1H$ atoms are replaced by $_3H$ atoms.

TRITIUM : A radioactive isotope of hydrogen having mass number 3. It has a half life of 12.3 years. It is used as a tracer in investigation work.

TRIVALENT : Having a valency of three.

TROPYLIUM IONS : The positive ion, $C_7H_7^+$, having a symmetrical seven membered ring of carbon atoms. It exhibits aromaticity.

TRYPSIN : An enzyme which can digest proteins. It is secreted by the pancreas.

TUNGSTEN : W; Wolfram; A white or grey coloured, transition element belonging to the 6th group and 6th period of the periodic table. Its atomic number is 74, while relative atomic mass is 183.85. Its important ores are wolframite $\{(FeMn)Wo_4\}$ and scheelite $(CaWO_4)$. It is used in making the filaments of electric bulbs.

TUNGSTEN CARBIDES : Very strong compounds, used to make tools and abrasives. They are made by heating powdered tungsten with carbon. WC is a high melting carbide and is a good conductor of electricity, while W_2C, having a similar melting point, is less conducting. It is very resistant to chemicals.

TURPENTINE : An oil extracted from pine resin. It contains terpene pinene, $C_{10}H_{16}$ and some other terpenes. It is used as a solvent.

TURQUOISE : A mineral consisting of hydrated phosphate of copper and aluminium, having composition $\{Cu\ Al_6\ (PO_4)_4\ (OH)_8 \cdot 4H_2O\}$. Generally blue in colour, it is used as a semi-precious stone.

TRYOSINE : An alpha amino acid constituent of proteins.

U

ULTRACENTRIFUGE : A method to seperate macromolecules like proteins or nucleic acids from solutions by high speed centrifuge. These are electrically operated.

ULTRAHIGH VACUUM : *See vacuum.*

ULTRA MICROSCOPE : A microscope to reveal the presence of particles that are invisible when seen with the optical microscope. It is based on Tyndall effect. In this, the particles produce diffraction ring systems, appearing as bright on the dark background.

ULTRAVIOLET : (UV); An electromagnetic radiation with wavelength Inm-400nm which is shorter than visible light. UV rays are also produced by electronic transitions between outer energy levels of atoms. UV photons carry more energy than that of light because of their higher frequencies.

UNCERTAININITY PRINCIPLE : Also called Heisenberg uncertarinity principle or principle of indeterminism.

It states that it is not possible to know the position and momentum of a particle with unlimited accuracy.

UNIMOLECULAR : It describes a reaction in which only one molecule is involved. The following are unimolecular reactions:

$N_2O_4 \rightarrow 2NO_2$

$H_2O_2 \rightarrow H_2O + {}^1/_2\ O_2$

UNIT : Reference value of a quantity which are used to express other values of the same quantity.

UNIT CELL : Smallest group of atoms, ions or molecules which, when repeated at regular intervals in 3D, will produce a crystal.

Lattice basic unit cells are seven in number and so seven crystal systems result.

UNIT PROCESSES : These are chemical conversions. These include steps in chemical reactions like distillation, nitration, alkylation, acytation, hydrogenation etc.

UNIVALENT : Also called monovalent. Having a valency of one.

UNIVERSAL INDICATOR : Also called multiple range indicator. A mixture of indicators that changes colour over a wide pH range. Typical indicators are methyl orange, phenolphthalein, and changes colour for pH equal to 3 to 10.

UNSATURATED COMPOUND : Organic compounds that contain one or more double or triple bond in its carbon chain. So multiple bonds are relatively weak. These compounds usually undergo addition reactions.

UNSATURATED SOLUTION : *See saturated solution.*

UNSATURATED VAPOUR : *See saturated vapour.*

URACIL : One of the major component bases of nucleotides RNA, which is a derivative of pyrimidine.

URANINITE : Mineral form of uranium (IV) oxide which has minute amounts of radium, thorium, lead and helium. It occurs in a massive form with a lustre in pitch-blende, which is one of the chief ores of uranium.

URANIUM : U; A white radioactive element of the actinide series of metals. It occurs in three naturally occuring radio isotopes ^{238}U, ^{238}U and ^{235}U is a major nuclear fuel.

URANIUM (VI) FLOURIDE : UF_6; Also called uranium hexaflouride. A volatile white crystalline solid which is used for separation of uranium isotopes.

URANIUM LEAD DATING : A method of dating certain rocks that depend on the decay of radio isotopes, U-238 to Pb-206 with a half life of 4.5×10^9 years, or the decay of U-235 to Pb-207 with a half life of 7.1×10^8 years. One method of measuring the age of rocks is to measure the ratio of radioactive lead to non-radiogenic lead (Pb^{-204}). The method gives reliable results for ages of the order 10^7-10^9 yrs.

URANIUM SERIES : *See radioactive series.*

Uranyl Derivatives : Compounds of U(VI) containing UO_2 grouping. Most uranyl derivatives are yellow in colour.

Urea : $CO(NH_2)_2$; Also called carbamide. MP 133°C. White crystalline compound made from ammonia and carbon dioxide. Also used in the manufacture of urea and formaldehyde series. Soluble in water, but insoluble in organic solvents.

Urea Cycle : Also called ornithine cycle. A series of biochemical reactions that converts ammonia to urea during excretion of metabolic nitrogen by mammals.

Urea formaldehyde Resin : Synthetic resins which are made by copolymerisation of urea with formaldehyde (methanal). Used as adhesives or thermosetting plastics.

Urethane Resins : Synthetic resins with group –NH–CO–O– as the repeating group in chain or polymers made by copolymerization of isocyanate esters with polyhydric alcohols. Used in plastics and paints.

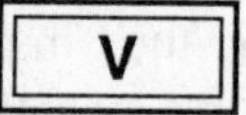

VACANCY : *See defect.*

VACUUM : Space containing no matter, but it is not possible to create that. Vacuum means a pressure below 10^{-2} pascal.

VACUUM DISTILLATION : The distillation of liquids under a reduced pressure. It helps in distilling the liquids much below their boiling points to avoid decomposition. Method is used for the purification and isolation of compounds, e.g., glycerol is purified by this method.

VACUUM PUMP : A pump used to reduce the gas pressure in an experiment. Different types of pumps can be used to maintain low pressures at different levels. Oil seal pump can maintain a pressure of 10^{-1} pa and diffusion pump can maintain a pressure of 10^{-7} pascal.

VALENCE : Valency; The combining capacity of an atom or radical, equal to the number of hydrogen atoms which may combine with an atom of the given element or double the number of oxygen atoms which can combine with the atom. For example valency of C in CH_4 is 4, of N in NH_3 is three. It is generally equal to the number of valence electrons, or eight minus the valence electrons.

VALENCE ELECTRON : Electron belonging to valence shell of an atom, capable of participating in chemical bond formation.

VALENCY : *See valence.*

VALERIC ACID : *See pentanoic acid.*

VALINE : An alpha amino acid.

VALIUM : Diazephas, a sedative.

VANADIUM : V; A silvery white transition metal belonging to the

5th group and 4th period, having atomic number 23 and relative atomic mass 50.94. It occurs in some complex ores like vanadinite and carnotite. It is used in a large number of alloy steels. It reacts with non-metals at high temperatures and shows oxidation states +5, +4, +3 and +2 in its compounds. In the beginning it was considered an impure form of chromium. It forms coloured compounds.

VANADIUM (V) OXIDE : V_2O_5; Vanadium pentoxide; An oxide used as catalyst in the contact process for the manufacture of sulphuric acid.

VAN-ARKEL PROCESS : A method for the purification of metals, e.g., Ti.

VAN DER WAAL'S EQUATION : An equation of state developed by removing the drawbacks in the postulates of the kinetic molecular theory of gases.

VAN DER WAAL'S FORCE : A weak intermolecular force of attraction arising from weak electrostatic interactions between molecules. These forces arise from dipole-dipole interaction, dipole-induced dipole interactions, disperson forces (resulting from temporary polarity arising from asymmetrical distribution of electrons). Such forces exist between the molecules of highly non-reactive substances like noble gases also.

VAN'T HOFF FACTOR : The ratio of the amount of observed colligative property and the calculated colligative property. It is related with the molecular masses also, as the ratio of molecular mass (normal)/molecular mass (observed).

VAN'T HOFF ISOCHORE : An equation relating the equilibrium constant of a reaction with temperature and enthalpy of reaction.

$$\frac{d\ln k}{dT} = \frac{dH}{RT^2}$$

VANT'S HOFF ISOTHERM : $\Delta G = -RT \ln k + RT \frac{[C]\cdot[D]}{[A]\ [B]}$

It relates free energy change with equilibrium constant at a constant temperature T.

VAPORISATION : The process of converting a liquid into its vapour phase by heating. It increases with the rise of temperature.

VAPOUR DENSITY : The ratio of the mass of a certain volume of vapour to the mass of an equal volume of hydrogen (measurements made under identical conditions). Molecular mass is twice the vapour density. It is determined by Victor Meyer's method or Duma's method.

VAPOUR PRESSURE : Pressure exerted by the vapour. At a particular temperature, it is defined as the pressure of the vapour in equilibrium with the liquid. It depends on the nature of the substance and the temperature.

VASELINE : Yellow or white mixture of hydrocarbons (alkanes having 15 to 20 carbon atoms). It is a semi-solid, used as a base for ointments.

VAT DYES : A class of insoluble dyes which are reduced to alkali soluble derivatives before application. After application, the insoluble dye is regenerated by atmospheric oxidation, e.g., indigo and indanthrene.

VERDIGRIS : The green layer made on the surface of copper and bronze exposed to the atmosphere. It is a mixture of basic copper carbonate and chloride.

VERMICULITE : A clay mineral with composition formula $NL(MgFe, Al)_3(AlSi)_4O_{10}(OH)_2 4H_2O$. Used as an insulating material and soil for pots.

VERMILIN : A pigment, mercury (II) sulphide.

VICINAL POSITIONS : Positions in a molecule at adjacent atoms, e.g., 1,2–dichloroethane.

VICTOR MAYER'S METHOD : A method for determining vapour densities and molecular masses of volatile liquids. In this method, a given weight of compound is vapourised and the volume of air displaced by it is measured. The volume of the displaced air is then reduced to NTP to calculate the vapour density and molecular mass.

VILLIAUMITE : A mineral form of sodium flouride.

VINEGAR : A dilute aqueous solution of ethanoic acid (4-10% acid) obtained by the oxidation of dilute aqueous solution of ethanol by air in presence of the micro–organism *Bacillus acetic* (Aceto bacter).

VINYL CHLORIDE : *See chloroethene.*

VINYL GROUP : $HC=CH_2$- group.

VIRIAL EQUATION : An equation representing the behaviour of real gases – $PV = RT + bP + cP^2 + dP^3 + ...$where b,c,d are called virial coefficients and the depend on the temperature.

VISCOSITY : The resistance to flow of a liquid. It is the reciprocal of fluidity. The measurements are used to find the molecular weights of polymers.

VISCOSITY, COEFFICIENT OF : n; It is the tangential force required per unit area to maintain a unit difference of velocity between two adjacent layers separated by a unit distance. Its units are poise (dynes sec cm^{-1}).

VITAMINS : Substances which are essential constituents of food. They are not present in proteins, carbohydrates, fats and mineral salts. They play a very significant role in animal metabolism. Deficiency of a particular vitamin in the food causes a definite disease called deficiency disease. Vitamins are of various types of which 14 have been generally recognised. Of these water soluble vitamins are B-complex and C, while fat soluble vitamins are A,D,E,K, etc.

VITAMIN A : Retinol ; A fat soluble vitamin. Present in green plants, animal fats, butter, yolk of egg and fish liver oils. Its deficiency causes night blindness, xerophthalmia and total blindness.

VITAMIN B–COMPLEX : A group of water soluble vitamins. Vitamin B_1 (thiamine) is present in brewer's yeast, wheat germ, beans, peas and green vegetables. Its deficiency causes beri–beri. Vitamin B_2 (riboflavin) occurs in green vegetables, yeast, liver and milk. Its deficiency causes inflammation of the tongue, lips etc. Vitamin B_6 (pyridoxine) is present in grains, yeast, liver, milk etc. Its deficiency causes retarding growth, dermatitis. Vitamin B_{12} is prepared by micro-organisms and its natural source is liver and is required for the normal production of red blood cells. Other vitamins in B-complex are nicotinic acid, pantothenic acid, folic acid etc.

VITAMIN C : Ascorbic acid; A colourless, crystalline, water soluble vitamin found in citrus fruits and green vegetables. It is needed for maintenance of normal health. Its deficiency leads to scurvy.

VITAMIN D : Fat soluble vitamin occuring in the form of two steroid derivatives D_2 (ergocalciferol and calciferol) and D_3 (cholecalciferol). These are formed by UV rays and sunlight respectively. Deficiency causes rickets in growing animals and osteomalacia in mature, i.e., deformity and weakness of bones.

VITAMIN E : Tocopherol; Fat soluble vitamin present in cereals, green vegetables etc. Its deficiency leads to several disorders in different species, including liver damage and infertility.

VITAMIN K : Fat soluble vitamins present in egg yolk and green vegetables, responsible for blood clotting. Its deficiency leads to extensive bleeding.

VITELLIN : Chief protein present in egg yolk.

VITREOUS : Having structure like glass.

VOLATILE : Describing a substance which is easily converted into vapours.

VOLHARD METHOD : A method for the estimation of Ag^+ using ions in the presence of Fe^{3+} ions acting as indicator.

VOLT : V; The S.I. unit of potential, potential difference and electromotive force. It is defined as the potential difference between two points in a circuit between which a constant current of one ampere flows when the power used is one watt. 1 volt = $1C^{-1}$.

VOLTAIC CELL : *See cell*

VOLTAMETER : Coulometer; An electrolytic cell previously used to measure the quantity of electric charge. It is based on Faraday's laws of electrodeposition. $W = ctz$, where ct = electricity in coulombs and z is the electrochemical equivalent of the metal.

VOLUME : V; The space occupied by a body or system. May be in any phase. Its units are m^{-3}.

VOLUMETRIC ANALYSIS : A method of quantitative analysis based on measurement of volumes. For liquids, it, involves titrations, e.g., acidimetery, alkalimetry, redox titrations, iodometry etc.

VULCANITE : Ebonite; A hard dark coloured insulating material made by the vulcanisation of rubber with a high proportion of sulphur (nearly 30%).

VULCANISATION : A process for hardening rubber when it is heated with sulphur or a suitable compound of sulphur.

W

WACKER PROCESS : An industrial method for making ethanal by the aerial oxidation of ethene in the presence of $PdCl_2$ and $CuCl_2$ as catalysts. Pd^{++} forms a complex with ethene (as intermediate), while $CuCl_2$ is used to oxidise the palladium back to its Pd^{++} state. Cu^{+} ions so formed, spontaneously undergo oxidation to Cu^{++} ions by air. The method is used to prepare ethanal as well as ethanoic acid.

WALDEN INVERSION : A reaction in which the configuration of an optically active compound gets inverted, keeping the compound still optically active. It happens in case of the SN2 reaction mechanism of hydrolysis of alkyl halides.

WARFARIN : 3–(alpha acetenoyl benzyl)-4-hydroxy cumarin; A synthetic coagulant. In large doses it acts as rodenticide.

WASHING SODA : $Na_2CO_3 . 10H_2O$.

WATER : H_2O; A colourless, odourless liquid at room temperature. In the gaseous phase, it exists in the form of single molecules, but in liquid and solid states a large number of molecules remain associated due to intermolecular hydrogen bonding. In ice there are some vacant spaces (cages) caused by hydrogen bonds which explains the lower density of ice as compared to water. Liquid water is a covalent compound, but polar OH bonds cause overall polarity in the molecule. Structure of a single molecule of water is angular with bond angle 104.5° due to the presence of two lone pairs of electrons on the oxygen atom. Pure water is very weakly ionised into hydronium (H_3O^+) ions and hydroxyl (OH^-) ions. This makes the basis for acid-base behaviour according to Arrhenius theory of electrolytic dissociation.

Water is a very good solvent. It is capable of dissolving ionic substances due to its polar nature, and organic substances are dissolved due to its tendency to make hydrogen bonds. Water serves as a medium for most chemical reactions. Biologically, it is the most important constituent of the living organisms.

WATER GAS : A mixture of carbon monoxide and hydrogen, made by passing steam over red hot coke. In earlier times, water gas was used as a source of hydrogen for Haber's process for the manufacture of ammonia. It can be used as a fuel also but the reaction producing it is endothermic and so its use as a fuel is not economical.

WATER GLASS : A viscous colloidal solution of sodium silicate in waters used to make silica gel. It is also used as a preservative.

WATER OF CRYSTALLISATION : Water molecules present in crystalline compounds in a definite ratio. The water molecules may simply occupy lattice positions in the crystal, or may form bonds with the metal ions. For example, in $CuSO_4.4H_2O$, four water molecules are linked to Cu^{++} ions through coordinate bonds forming complex cation $[Cu(H_2O)^4]^{2+}$. Fifth water molecule is held by SO_4^- through hydrogen bonding.

WATER SOFTENING : *See hardness of water.*

WATT : W; S.I. unit of power. It is defined as a power of one joule per second. In electricity, it is defined as rate of energy change by an electric current of one ampere flowing through a conductor, the ends of which are maintained at a potential difference of one volt. It is named after James Watt.

WAVE FUNCTION : It is equal to the amplitude of an electron wave in given a orbital. The physical significance of wave function is that its square at a point is proportional to the probability of finding the particle in a small volume around the nucleus. For an electron in an atom, it gives rise to the idea of atomic orbital.

WAVE-LENGTH : λ; The distance between the ends of one complete cycle of a wave or the distance between two neighbouring

crests or troughs. It is mathematically given as $\lambda = \frac{c}{f}$, where f is the frequency of radition. For matter waves, it is given by de Broglie's equation, $\lambda = h/p$, where p is the momentum of the particle.

WAVE MECHANICS : *See quantum mechanics.*

WAVE NUMBER : Reciprocal of the wavelength. It is the number of waves formed in unit distance. Its unit is cm^{-1} or m^{-1}

WAX : Soft amorphous, water repellent organic solid. The term is generally used for the esters of long chain monohydric alcohols with fatty acids. Waxes secreted by plants and animals have a protective function.

WEAK ACID : An acid which is only slightly ionised, e.g., acetic acid.

WESTON CELL : A type of primary galvanic cell used as a standard cell, producing a constant e.m.f. of 1.0186 V at 20°C. The H - shaped apparatus, in one arm, has a paste of cadmium sulphate and mercury(I)sulphate, while in the other arm it has cadmium amalgam cathode covered with cadmium sulphate. Electrolyte in both the arms is sataurated soluion of cadmium sulphate.

WESTRON : $Cl_2CHCHCl_2$; 1,1,2,2,-tetrachloroethane made by the complete chlorination of ethyne. It is used as an industrial solvent for rubber, fats and varnishes. It also has insecticidal action.

WESTROSOL : $Cl_2C=CHCl$; 1,1,2;-trichloroethene; Obtained by passing westeroso vapours over heated barium chloride or lime. It may also be made by the action of a limited amount of alcoholic potash on westroso. It is also a good solvent.

WHITE ARSENIC : *See arsenic (III) oxide.*

WHITE LEAD : *See lead(II)carbonate hydroxide.*

WHITE MILA : *See muscovite.*

WHITE SPIRIT : A solvent for paint. It is liquid mixture of hydrocarbons obtained from petroleum.

WILLIAMSON'S SYNTHESIS : A method for preparing simple and mixed ethers by the action of alkyl halide and sodium alkoxide in an alcoholic medium. When a mixed or unsymmetrical ether is being made, a possible side reaction is eliminations which yields an alkene and an alcohol.

There is another method called Williamson's continuous process, in which an alcohol is dehydrated by concentrated sulphuric acid when alcohol vapours are passed through at a temperature of 413K. This method can be used for making only symmetrical or simple ethers.

WITHERITE : A mineral form of barium carbonate.

WITTIG'S REACTION : A reaction between an alkylidence phosphorane (ylide) and an aldehyde or a ketone to form an alkene.

WOHLER'S SYNTHESIS : Synthesis of urea by heating and evaporating a solution of ammonium isocyanate. This synthesis disputed the original theory that organic compounds were made only by living organisms. It was carried out by the German chemist Friedrich Wohier.

WOLFRAM : *See tungsten.*

WOLFRAMITE : A mineral containing iron, manganese and tungsten, having a composition $(FeMn)WO_4$. Black or brown crystalline solid is the principal ore of tungsten.

WOOD'S METAL : A low melting alloy of 25% lead, 50% bismuth and 12.5% each of tin and cadmium. It is used for fusible links in sprinklers.

WOOD SPIRIT : *See methanol.*

WORK FUNCTION : A quantity which determines the extent of photoelectric emission according to Einstein's photoelectric equation. It may be expressed as potential difference in volts (with function potential) or as the work to be done by the excited electron in electron volts or joules (work function energy).

WROUGHT IRON : A highly pure form of iron, containing negligibly small amount of carbon and nearly 1-3% slag (iron silicate). Wrought iron rusts less easily. It is used for making hooks, tubes, chains etc.

WURTZITE : A mineral form of zinc sulphide.

WURTZ REACTION : A method of preparing alkanes by refluxing a halo alkane with sodium in dry ether. The method is used to prepare symmetrical alkanes. The method is modified to get alkyl benzenes by refluxing alkyl halide, aryl halide and sodium in dry ether and is called Wurtz Fitting reaction.

But this method gives other products also, for example, ethane will also be formed.

XANTHATES : Salts or esters containing the group – SCS(OR) where R is an alkyl or some other organic group. Cellulose xanthate is an intermediate in the manufacture of rayon. Xanthates are used in curing and vulcanising rubber.

XANTHANE : $C_3H_8O_2$; A parent of xanthone group of dyes. It is made by heating phenyl salicylate.

XENON : Xe; A colourless, odourless gas belonging to the 18th group and 5th period of the periodic table. Its atomic number is 54, while its relative atomic mass is 131.30. It is present in the atmosphere in a very small amount and is obtained by the fractional distillation of liquid air. The element is used in fluorescent lamps. Although a noble inert gas, it is found to combine with fluorine and oxygen to form several compounds. The credit for the discovery of its compounds goes to N. Bartlett. Its important compounds are $XePtF_6$, DeF_2, XeF_4, $XeSiF_6$, XeO_3, XeO_2F_2 The element was discovered by Ramsay and Travers.

XEROGRAPHY : A process in which a photoconductor on surface is charged, exposed to the image, which discharges all except the image areas, removal of photoconductor from non image areas, developing the image with an oppositely charged pigment, transferring the image to paper electrostatically and fixing the image on paper by warming.

X-RAY CRYSTALLOGRAPHY : The technique involves directing X-rays at a crystalline sample and recording the diffracted X-rays on a photographic plate. Crystal structure can be worked out from the positions and intensities of the spots on the diffraction pattern.

X-RAY DIFFRACTON : The wavelengths of X-rays are comparable to the distances between the atoms of most of the crystals. A crystal of suitable type can be used to disperse X-rays in a spectrometer. It is also the basis of X-ray crystallography.

X-RAY FLUORESCENCE : The emission of X-rays from excited atoms produced by the collisions of high energy electrons and other particles. The wavelengths of the fluorescent X-rays can be measured by X-ray spectrometer. It is used in electron-probe microanalysis.

X-RAYS : Electromagnetic radiations of wavelengths shorter than UV radiations, but longer than gamma radiations. Atoms of all the elements emit a characteristic X-ray spectrum when they are bombarded by electrons. X-rays can pass through many forms of matter and so they are used medically and industrially to examine internal structures. The rays are further used in crystallography and radiography.

XYLENES : C_8H_{10}; Dimethyl benzenes. Exist in three forms, viz., o, m and p–xylenes, obtained by reforming naphthalene in the presence of hydrogen.

XYLIC ACID : 2,4-dimethyl benzoic acid.

Y

YEASTS : A group of fungi which are used in baking and alcohol industry. Yeasts contain enzymes which function as biocatalysts.

YLIDES : $R_2C=P(C_6H_5)_3$ or $R_2C^- P^+(C_6H_5)_3$, Made by the action of alkylhalide and triphenyl phosphide. These are important compounds which react with carbonyl compounds to make alkenes, e.g., benzaldehyde reacts with a ylide $CH_2=P(C_6H_5)_3$ and produces styrene.

YTTERBIUM : Yb, A silvery, inner transition metal, having atomic number 70 and relative atomic mass 173.04. It is a lanthanide and so occurs with the other lanthanides. It mainly occurs in gadolinite monazyte. It was discovered by J.D.G. Marignac.

YTTRIUM : Y, A silvery grey transition metal of the 3rd group and 5th period of the periodic table. Its atomic number is 39 and relative atomic mass in 88.905. It occurs in uranium ores and lanthanide ores. Metal is used in super conductor alloys, and for making permanent magnets. Its oxide Y_2O_3 is used in colour television phosphors. Chemically, it resembles lanthanides. It was discovered by Friedrich Wohler.

Z

Zeeman Effect : Splitting of spectral lines under the influence of a magnetic field. If the magnetic field is parallel to the light path, each line splits into two lines, and into three when the field applied is perpendicular to the path of light.

Zeisel Reaction : The reaction of an ether with excess concentrated hydroiodic acid to form a mixture of alkyl iodides. It helps in establising the structure of the original ether Zėise's salt.

Zeolite : A group of hydrated aluminosilicate minerals which occur in nature. They are used as water softners and for the refining of sugar.

Zero Order : Describing a chemical reaction in which the rate of reaction is independent of the concentration of reactants, e.g., reaction between hydrogen and chlorine.

Zero Point Energy : The energy possessed by the atoms and molecules of a substance at zero kelvin.

Zeroth Law Of Thermodynamics : *See thermodynamics.*

Ziegler Natta Catalyst : Solution of titanium chloride and trialkyl aluminium. It is used as a catalyst for the polymerisation of alkenes.

Ziegler's Process : A method for the manufacture of high density polyethene using Ziegler Natta catalyst ($TiCl_4$ + Al $(C_2H_5)_3$).

Zinc : Zn; A transition metal having atomic number 30 and relative atomic mass 65.38. It belongs to the 12th group and 4th period of the periodic table. Its properties do not completely resemble those of transition elements due to completely filled (n-1) d subshell. It occurs in nature as zinc blende (ZnS) and Smithsonite ($ZnCO_3$).

ZINC BLENDE : A mineral form of zinc sulphide, ZnS, an important ore of zinc. In zinc blende, zinc atoms are surrounded by lone sulphur atoms, each at the corner of a tetrahedron. Each sulphur is also surrounded by four zinc atoms. These crystals belong to the cubic system.

ZINC CHLORIDE : $ZnCl_2$; A white crystalline substance previously called butter of zinc. The anhydrous salt is deliquescent and may be made by the action of hydrogen chloride gas on hot zinc. It exists in various hydrated forms also. Used in dentistry, in Leclanche cells, as a catalyst and as a dehydrating agent.

ZINC GROUP : The group of elements, Zn Cd and Hg, in the periodic table. It is considered as the last group of transition elements.

ZINCITE : A mineral form of zinc oxide.

ZINC OXIDE : ZnO; A white powder made by oxidising hot zinc in air or by the thermal decomposition of zinc nitrate or zinc carbonate.

ZINC SULPHATE : $ZnSO_4$; A white crystalline solid, made by heating ZnS in air or by the action of dilute sulphuric acid on zinc oxide or zinc carbonate. Its heptahydrate is called white vitriol. On heating, becomes a hydron at 723K. It is used in the textile industry as mordant.

ZINC SULPHIDE : ZnS; A yellowish white, water soluble compound. It occurs in nature as sphalerite and wurtzite. It may be prepared by passing H_2S gas through an ammonical solution of zinc salt. Impure zinc sulphide is phosphorescent. It is used as a pigment and to make luminescent screens.

ZIRCONIA : ZrO_2; Zirconium (IV) oxide. Used as electrolyte in fuel cells.

ZIRCONIUM : Zr; A white transition metal that occurs in nature in a gemstone, zircon Zr SiO_4, and in baddeleyite ZrO_2. It has atomic number 40 and relative atomic mass 91.22 It belongs to the 4th group and 5th period of the periodic table. The element is used in nuclear reactors (as an effective neutron absorber) and in certain alloys. The metai was discovered by Klaproth.

ZWITTERION : Internal salt, dipolar ion or ampholyte ion. It is an ion that has +ve and -ve charges on the same group of atoms.